# Youth in a Suspect Society

# YOUTH IN A SUSPECT SOCIETY: DEMOCRACY OR DISPOSABILITY?

Henry A. Giroux

palgrave
macmillan

First published in hardcover in 2009 by PALGRAVE MACMILLAN®
in the United States - a division of St. Martin's Press LLC,
175 Fifth Avenue, New York, NY 10010.

Where this book is distributed in the UK, Europe and the rest of
the world, this is by Palgrave Macmillan, a division of Macmillan
Publishers Limited, registered in England, company number 785998,
of Houndmills, Basingstoke, Hampshire RG21 6XS.

Palgrave Macmillan is the global academic imprint of the above
companies and has companies and representatives throughout the world.

Palgrave® and Macmillan® are registered trademarks in the United
States, the United Kingdom, Europe and other countries.

ISBN: 978–0–230–10870–7

Library of Congress Cataloging-in-Publication Data

Giroux, Henry A.
    Youth in a suspect society : democracy or disposability? /
    Henry A. Giroux.
        p.   cm.
    Includes bibliographical references and index.
    ISBN 0–230–61329–2
    1. Youth—United States—Social conditions—21st century.
    I. Title.
    HQ796.G527 2009
    305.2350973′09051—dc22                          2009006671

Design by Integra Software Services

First PALGRAVE MACMILLAN paperback edition: November 2010

10 9 8 7 6 5 4 3 2 1

Printed in the United States of America.

To the Memory of my nephew: Alfred U. Barbery III
September 2, 1967–April 18, 2008

To the Memory of my comrade and colleague: Joe L. Kincheloe
December 14, 1950–December 19, 2008

# CONTENTS

# ACKNOWLEDGMENTS

This book was written at a point in history marked by a combination of deep despair and soaring hopes. As the counterrevolution that has gripped the United States since the late 1980s appears to be coming to an end with the election of Barack Obama, the dark times that befell us under the second Bush administration no longer prompt either unrelenting despair or collective cynicism. The assault that the second Bush administration waged on practically every vestige of the public good—from the Constitution to the environment to public education—appears to have lessened its grip as the country prepares for a new administration. Yet the range, degree, and severity of the problems the Obama team will inherit from the Bush administration seem almost too daunting to address: a war raging in two countries, a legacy of torture and secret prisons, a dismantling of the regulatory apparatus, a poisonous inequality that allocates resources to the rich and misery to the poor, an imperial presidency that has shredded the balance of power, a looming ecological apocalypse, a ruined reputation abroad, and a financial crisis that is almost unprecedented in American history—policies and conditions that have brought great suffering to millions of Americans and many millions more throughout the world. But there is more at stake here than an economic crisis. As Chris Hedges argues, accompanying the financial meltdown is the emergence of corporate public relations machine endlessly appealing to a cult of the self "elaborately constructed by the architects of our consumer society, which dismisses compassion, sacrifice for the less fortunate and honesty." This pernicious mode of public pedagogy is consistently reinforced by "reality television programs, business schools and self-help gurus....[who claim that] success, always defined in terms of money and power, is its own justification. The capacity for manipulation is what is most highly prized." What we are left with is a "moral collapse [that] is as terrifying, and as dangerous, as our economic collapse."[1] But the economic and moral crisis that is most often forgotten or repressed in the daily headlines of gloom is the war that is being waged at home, primarily against young people, who have historically been linked to the promise of a better life, one that they would both inherit and reproduce for future generations. In a radical free-market culture, when hope is precarious and bound to commodities and a corrupt financial system, young people are no longer at risk: they are the risk. The conditions produced by the financial crisis have resulted in the foreclosure of not only millions of family homes but also the future of young people, as the prospects of the unborn are mortgaged off in the interests of corporate power and profits. As wealth has moved furiously upward into private hands for the last several decades,[2] any talk about the future has had less to do with young people than with short-term investments, quick turnovers in profits, and the dismantling of the welfare state. Moreover, the destruction of the welfare state has gone hand in hand with the emergence of a prison-industrial complex and a new carceral state that regulates, controls, contains, and punishes those who are not privileged by the benefits of class, color, immigration status, and gender. How else to explain a national prison population that has grown from 200,000 in 1973 to slightly over 2.3 million in 2008? It gets worse. The Bureau of Justice Statistics reports that at the end of 2007 "over 7.3 million people were on probation, in jail or prison, or on parole—3.2% of all U.S. adult residents or 1 in every 31 adults."[3]

As policing, containment, and imprisonment merge with a market-driven society that places both the reasons for and redress of misfortune entirely in the hands of isolated individuals, the circuitry of social control redefines the meaning of youth, subjecting particularly those marginalized by class and color to a number of indiscriminate, cruel, and potentially illegal practices by the criminal justice system. In the age of instant credit and quick profits, human life is reduced to just another commodity to be bought and sold, and the logic of short-term investments undercuts long-term investments in public welfare, young people, and a democratic future. Not surprisingly, youth as a symbol of long-term commitment are now viewed as a liability rather than an asset. Barack Obama ran on a platform that redefines both the future and the promise of a democracy to live up to its obligations to future generations. The election of Obama to the presidency signaled a demand for change and a collective call for a new kind of politics, one that hopefully will reclaim and act on the democratic ideals that have been progressively subverted during the last three decades. One indeed hopes that young people will figure prominently in that promise.

This book analyzes the forces that ushered in such dark times and examines their most unlikely and often invisible victims—those young people who now symbolize trouble rather than promise and who acutely feel the repercussions of adult neglect, if not scorn, especially those youth for whom race and class loom large in their lives. This is a generation of young people who have been betrayed by the irresponsibility of their elders and relegated to the margins of society, often in ways that suggest that they are an excess, a population who, in the age of rampant greed and rabid individualism, appear to be expendable and disposable. Moreover, *Youth in a Suspect Society* attempts to construct a new theoretical discourse for addressing both the suffering many young people experience, albeit to different degrees, and the promise that a revitalized democracy might offer them. In many ways, this book is motivated by a sense of outrage and a sense of hope. On the one hand, it identifies a number of forces—including unfettered free-market ideology, a dehumanizing mode of consumerism, the rise of the racially skewed punishing state, and the attack on public and higher education—that have come together to pose a threat to young people, forces so extreme they can be accurately described as a "war on youth." On the other hand, in making visible those forces responsible for such an attack, this book also points to practices and conditions necessary to challenge and overcome the dire state of today's youth. At the heart of this intervention is the belief that individual and collective resistance is born out of awareness, education, good judgment, and an ethic of mutuality—all of which suggests a struggle that is as educational as it is political, with no line dividing one from the other. While there are good reasons to celebrate the Obama victory, it offers no guarantees that the political, economic, and social conditions that have brought us to the brink of disaster will fundamentally change. Substantive and lasting change must come from below: from young people, students, workers, intellectuals, artists, academics, parents, and others willing not just to demonstrate for equality, freedom, and social justice but to organize in order to push hope over the tipping point, push politics in a new direction, and engage in a collective struggle that takes power away from political and corporate elites, returning it to the people who are the real source of any viable democracy.

*Youth in a Suspect Society* addresses the changing conditions youth now face in the new millennium and the degree to which they have been put at risk by social policy, institutional mismanagement, and shifting cultural attitudes. While youth have always represented an ambiguous category, they have within the last 30 years been under assault in ways that are entirely new, and they now face a world that is far more dangerous than at any other time in recent history. This book develops a new set of categories and vocabulary for understanding the changing conditions of youth within the relentless expansion of a global market society, one that punishes all youth by treating them largely as commodities. It also explores a much darker side of radical free-market ideology, with its emphasis on deregulation, privatization, and the elimination of any vestige of the social state, one that subjects poor youth and youth of color to the harshest elements, values, and dictates of a growing youth-crime complex, governing them through a logic of punishment, surveillance, and control. In this instance, even as the corporate state is in turmoil, it is transformed into a punishing state, and certain segments of the youth population become the object of a new mode of governance based on the crudest forms of disciplinary control.

Any discourse about youth out of ethical necessity should raise serious questions about the social and political responsibility of educators in addressing the plight of young people today. What is the purpose of higher education and its faculties in light of the current assault on young people, especially since it is education that provides the intellectual foundation and values for young people to understand, interrogate, and transform when necessary the world in which they live? Matters of popular consciousness, public sentiment, and individual and social agency are far too important as part of a larger political and educational struggle not be taken seriously by academics who advocate the long and difficult project of democratic reform. Tragically, few intellectuals providing critical commentary on the current tragic conditions affecting youth offer any insights regarding how the educational force of the culture actually works to reproduce dominant ideologies, values, identifications, and consent. That is, there are too few commentaries about how the media, schools, and other educational sites in the culture provide the ideas, values, and ideologies that legitimate the conditions that enable young people to become either commodified, criminalized, or made disposable. While it is important to reform the economic and political structures that oppress young people, it is not enough. Yet, how exactly would it be possible to imagine a more just, more equitable transformation in government and economics without a simultaneous transformation in culture, consciousness, social identities, and values? *Youth in a Suspect Society* considers the role that academics and institutions of higher education may take in addressing the crisis of youth and its relationship to politics and critical education. Finally, it is impossible to understand the current crisis of youth and democracy without situating such a crisis in a larger theoretical and historical context. In addressing this challenge, I provide a broader analysis of what I call the biopolitics of neoliberalism and disposability, examining it not only as an economic system, but also as an educational, cultural, and political discourse that has laid the groundwork for a set of practices and policies in which young people are increasingly defined through market-driven ideas, social relations, and values that are predatory in nature and punishing in their consequences, leaving a generation of young people with damaged lives, impoverished spirits, and bankrupted hopes. Only by understanding the pervasive and all-embracing reach of neoliberalism with its discredited Gilded Age politics and values does it become possible to grasp the contours of a new historical period in which a war is being waged against youth, one that offers no apologies because it is too arrogant to imagine any resistance, even in the face of a financial and economic meltdown. Fortunately, power is never completely on the side of domination; nor is it entirely in the hands of those who view youth as an excess to be contained or burden to be expelled. Power is also born of a realistic sense of hope, one that situates new possibilities and dreams of the future within the realities of current structures of domination and oppression. Young people have always been a beacon for such hopes and now they have an opportunity to become, once again, agents of real change. Young people are on the cusp of a historic moment, and this book is composed in solidarity with them and the next generation of youth who hopefully will never allow this terrible assault on democracy to recur.

This book would not have been completed on time without the help of many friends who offered invaluable criticisms and support. I would like to especially thank David Clark, Sophia McClennen, Nasrin Ramihieh, Christopher Robbins, Ken Saltman, Cary Fraser, John Comaroff, Zygmunt Bauman, Howard Zinn, Roger Simon, Donaldo Macedo, Nick Couldry, Stanley Aronowitz, Doug Kellner, Carol Becker, Toby Miller, Peter Mayo, David Theo Goldberg, Lawrence Grossberg, Brian McKenna, Lynne Worsham, Olivia Ward, Adam Fletcher, Tony Kashani, Michael Peters, Dean Birkenkamp, and Doug Morris. Thanks also to Reno for his generosity and wonderful conversations and to Dr. Lawrence Hart and Dr. Bruce Korman for service beyond the call of duty. A special thanks to Subhash G. Dighe, whose smile and sharp wit are only matched by his miraculous ability to heal. Also, thanks to Scott Huen for his charm, generosity, and life saving skills. The notion of "The Suspect Society" I have borrowed from Mary O'Connell, who produced for CBC radio a three-part program using the concept to name her investigation of the rise of authoritarian elements in the United States. Heartfelt love to my sister Linda Barbery and my brother-in-law Al Barbery. And, as always, thanks to my three boys, Jack, Chris, and Brett, for bringing so much joy into my life. I am deeply indebted to Grace Pollock, my research assistant and colleague, who provided detailed editorial suggestions on the entire manuscript, was a great source of ideas, and greatly improved the quality of the overall project. I also want to think Maya Sabados, my administrative assistant, whose support was as outstanding as it was professional. Her skills and help went far beyond any standard measure of excellence. She typed notes, collected research, read chapters, and did everything with enormous focus and grace. She is truly one of my guardian angels. I also want to thank Julia Cohen, a brilliant poet who also happens to be a terrific editor, for supporting this project and putting up with all of my missteps. The love of my life, Susan Searls Giroux, provided a number of insightful and critical ideas, put up with endless queries, dazzled me with her conversations and choice of red wines, and read with great care every word in the manuscript. This book would simply not have been written without her critical interventions, ideas, and ever-appreciated presence. I am delighted that Kaya, my canine companion, slept under my desk as I wrote most of these chapters—providing love, warmth, and an antidote to my often-escalating blood pressure. And a warm hug to our new canine companion, Miles, who cheerfully slept at my feet while I edited the manuscript, offering a smile and a bark whenever he sensed I was getting restless.

I am dedicating this book to two wonderful human beings who left the world much too early and whose passing affects a great many people. My friend of 30 years, Joe Kincheloe, passed away unexpectedly, leaving us numb with grief and saddened that we did not have more time to spend with him. Joe offered a rare combination of gentleness and profound passion. He was brilliant, energetic—often on fire— and always a pleasure to be around. His never-flagging enthusiasm for critical pedagogy and social justice manifested itself in a number of important books, his unfailing support of countless students, and his always enjoyable presence in the lives of those with whom he shared ideas, love of music, and company. This book is dedicated to his memory. This book is also dedicated to my nephew, Alfred U. Barbery III, a devoted public servant, father, and son. He was much too young to leave this life and his death has brought great sorrow to many people who loved him and will miss him terribly. This book is an affirmation of admiration of his life and memory in the midst of the unimaginable pain shared by our family.

A version of chapter 4 was published in "Beyond the Biopolitics of Disposability: Rethinking Neoliberalism in the New Gilded Age," *Social Identities* 14:5 (September 2008), pp. 587–620.

## NOTES

1. Chris Hedges, "America is in Need of a Moral Bailout," *Truthdig* (March 23, 2009. Online: http://www.truthdig.com/report/item/20090323_america_is_in_need_of_a_moral_bailout/)
2. See, for example, Dollars and Sense and United for a Fair Economy, *The Wealth Inequality Reader*, second edition (Boston: Dollars and Sense, 2008).
3. U.S. Department of Justice, "Bureau of Justice Statistics," accessed January 2008. Online: http://www.ojp.usdoj.gov/bjs/pandp.htm.

# INTRODUCTION

# EXPENDABLE FUTURES: YOUTH AND DEMOCRACY AT RISK

> *There is a growing consciousness of children at risk. But . . . there is also a growing sense of children themselves as the risk—and thus of some children as people out of place and excess populations to be eliminated, while others must be controlled, reshaped, and harnessed to changing social ends. Hence, the centrality of children, both as symbolic figures and as objects of contested forms of socialization in the contemporary politics of culture.*
>
> —Sharon Stephens, *"Children and the Politics of Culture in 'Late Capitalism'"*[1]

In spite of the almost unprecedented financial and credit crises gripping the United States, the legacy of jaded excess lives on as both a haunting memory and an ideological register that continues to shape contemporary politics. After all, it was only a few years ago that it was widely recognized, if not celebrated, that the New Gilded Age and its updated " 'dreamworlds' of consumption, property, and power" had returned to the United States with a vengeance.[2] The exorbitantly rich along with their conservative ideologues publicly invoked and celebrated the period in nineteenth-century American history when corporations ruled political, economic, and social life and an allegedly heroic entrepreneurial spirit brought great wealth and prosperity to the rest of the country. Even the *New York Times* ran a story in the summer of 2007 that contained not only a welcome endorsement of Gilded Age greed but also praise for a growing class of outrageously rich chief executives, financiers, and entrepreneurs, described as "having a flair for business, successfully [breaking] through the stultifying constraints that flowed from the New Deal" and using "their successes and their philanthropy [to make] government less important than it once was."[3] But there was more at work in these examples of Gilded Age excess than a predatory narcissism, a zany hubris, and a neofeudal worldview in which self-interest and the laws of the market were seen as the only true measure of politics. There was also an attack on the idea of the social contract—in which the state gives minimum guarantees of security—and the very notion of democratic politics. In this second and now humbled Gilded Age, people were not bound together as citizens but as consumers, while the neoliberal values of self-interest, personal advancement, and economic calculation rendered ornamental "the basic principles of and institutions of democracy."[4]

As people in the United States struggle with the deepening financial and credit crisis that is wreaking havoc with their lives, it is important to recognize that the second Gilded Age with its rampant corruption, market deregulation, and unparalleled concentrations of wealth is both the outgrowth and the backdrop of what I define here as a radical mode of economic Darwinism and unfettered free-market values. And it is precisely this combination of market fundamentalism, greed, and cutthroat individual competition that has produced an unparalleled degree of social inequality and massive dislocations in the basic foundations of the larger society. Clearly, any understanding of the present financial crisis and its disastrous effects on young people has to be understood in terms of Gilded Age excess and the ideologies, cultural formations, economic forces, and modes of political irresponsibility that produced it.

Under free-market fundamentalism, or neoliberalism as it is called in some quarters, social problems become utterly privatized and removed from public considerations. Principles of communal responsibility are derided in favor of individual happiness, largely measured through the acquisition and disposability of consumer goods. In this highly privatized universe, visions of the good society are cast aside, replaced "by the perpetual search for bargains"[5] and maximum consumer satisfaction. The consequences involve not only the undoing of the social bond and importance of shared responsibilities, but also the endless reproduction of much-narrowed registers of character and individual self-reliance as a substitute for any analyses of the politics, ideologies, and mechanisms of power at work in the construction of socially created problems. This makes it more socially acceptable to blame the poor, homeless, uninsured, jobless, and other disadvantaged individuals and groups for their problems, while reinforcing the merging of the market state with the punishing state. The consequences of coupling market fundamentalism and a new authoritarianism have had a particularly devastating effect on youth in the United States, especially those who are marginalized by class and race. No longer seen as a social investment or the central element of an increasingly embattled social contract, youth are now viewed as either consumers, on the one hand, or as troubling, reckless, and dangerous persons, on the other. This assault against youth—and the long-term effects of such an assault on the possibility of a democratic future itself—should not be underestimated, nor its implications for countries under the dominating influence of American power. As the financial crisis looms large in the lives of the majority of Americans, government funds are used to bail out the banks and transnational corporations rather than being used to address the growing impoverishment of those who have lost homes, jobs, and any hope of a better future. As the crisis unfolds, it appears unlikely that a change of government in 2009 can undo the damage done by the Bush administration (itself the culmination of a century-long trend toward market deregulation) or reverse the effects of a rampant neoliberalism now unleashed across the globe.

The havoc wreaked by neoliberal economic policies can be seen in the hard currency of human suffering such policies have imposed on children, readily evident in some astounding statistics that suggest a profound moral and political contradiction at the heart of one of the richest democracies in the world. For example, the rate of child poverty rose in 2004 to 17.6 percent, pushing upward the number of poor children to 12.9 million. In fact, "[a]bout one in three severely poor people are under age 17."[6] Moreover, children make up a disproportionate share of the poor in the United States in that "they are 26 per cent of the total population, but constitute 39 per cent of the poor."[7] As a result of the severe economic crisis, Dr. Irwin Redlener, President of the Children's Health Fund in New York, claims that the number of children in poverty may increase to 17 million by the end of 2009.[8] Just as alarmingly, 9.3 million children lack health insurance, and millions lack affordable child care and decent early childhood education. One of the most damaging statistics revealing how low a priority children are in America can be seen in the fact that among the industrialized nations in the world the United States ranks first in billionaires and in defense expenditures and yet ranks an appalling twenty-fifth in infant mortality. As we might expect, behind these grave statistics lies a series of decisions to favor economically those already advantaged at the expense of youth. Moreover, for the last three decades we have witnessed, especially under the second Bush administration, savage cuts to education, nutritional assistance for impoverished mothers, veterans' medical care, and basic scientific research, all of which helped fund tax cuts for the inordinately rich. It also seems reasonable to assume that under the current financial crisis, young people, particularly youth marginalized by class or color, will experience even greater economic and educational hardships, while becoming even more invisible to the larger society.

## NEOLIBERALISM AS A THEATER OF CRUELTY

Under the George W. Bush administration, Americans spent almost a decade watching the dismantling of the social state, a decline in the use of its meager government provisions, radical deregulation, and the transfer of public wealth into the private bank accounts of the upper echelons of society. Rather than expanding the realm of freedom, neoliberal economics has brought us the financial crisis of 2008 along with an egregious number of housing foreclosures, a ruinous increase in personal bankruptcies, the loss of millions of jobs, and devastating cuts in state budgets and social services, all of which add up to an exemplary case of greedy financial markets out of control.[9] As Peter Dreier points out, "Wealth [became] even more concentrated during the Bush years. Today, the richest one percent of Americans has 22 percent of all income and about 40 percent of all wealth. This is the biggest concentration of income and wealth since 1928."[10] Most people around the world are aware of the precipitous decline of democracy in the United States, and there is a general global consensus that the domestic and foreign policies put into place since 2001 rightly qualify the George W. Bush administration as, in the words of former president Jimmy Carter, "the worst in history."[11] In fact, Carter's assessment of the Bush II regime seems tame compared to comments made over the last few decades by writers as renowned as Robert Kennedy, Jr., Seymour M. Hersh, and Gore Vidal, all of whom have argued that the United States has displayed the earmarks of an authoritarian regime.

After September 11, 2001, the United States moved even more rapidly away from a liberal democracy toward a punishing society. Bush's policies nourished and strengthened a number of antidemocratic forces, fostering a distinctive type of authoritarianism in the United States, including the militarization of everyday life, an imperial presidency, the use of state-sanctioned torture, the rise and influence of right-wing Christian extremists, and a government draped in secrecy that was all too willing to suspend civil liberties.[12] This emergent authoritarianism was largely legitimated through an ongoing culture of fear and a form of patriotic correctness designed to bolster a chauvinistic nationalism and a selective popularism. Fear was mobilized through both the war on terrorism and "the sovereign pronouncement of a 'state of emergency' [that generated] a wild zone of power, barbaric and violent, operating without democratic oversight in order to combat an 'enemy' that threatens the existence of not merely and not mainly its citizens, but its sovereignty."[13] As Stanley Aronowitz points out, the national security state was organized through "a combination of internal terrorism and the threat of external terrorism," which worked to reinforce "its most repressive functions."[14]

Within these expanding spheres of insecurity, privatization, deregulation, outsourcing, and a marauding market fundamentalism, the primary political and economic forces shaping American life indicated what was unique to the Bush version of neoliberalism: its hatred of democracy and dissent. As finance capital reigned supreme over American society, bolstered by the "new and peculiar power of the information revolution in its electronic forms,"[15] democratization along with the public spheres needed to sustain it became an unsettled and increasingly fragile, if not dysfunctional, project. Moreover, opposition to U.S. domination and the Bush administration was now answered not with the rule of law, however illegitimate, but with the threat or actuality of violence.[16] Hence, it is not surprising that the war at home gave rise to a crushing attack on civil liberties. This was evident in the passing of the Military Commissions Act of 2006, which conveniently allows the government to detain indefinitely anyone deemed an "enemy combatant" while denying recourse to the traditional right to challenge his or her detention through legal means. It was also apparent in the ongoing assault on those populations considered disposable and redundant under the logic of a ruthless market fundamentalism.

While the United States has never been free of repression, there was a special viciousness that marked the George W. Bush regime. The celebration of war, a landscape of officially sanctioned violence against people of color, a predatory culture of fear, and an attack on human rights coupled with the assault on the social state and the rise of an all-encompassing militarism make the Bush administration stand out in history for its antidemocratic policies. Yet even in the aftermath of the October 2008 global financial crisis and the historic election of Barack Obama as the forty-fourth president of the United States, the vocabulary and influence of corporate power and hapless governance can still be heard as the United States continues, albeit more slowly, along the trajectory of privileging corporate interests over the needs of the public good and the rising demands of millions of people struggling for economic, racial, and political justice. The shameless use of government power to reproduce corporate power and right-wing ideological interests was on full display as the Bush administration scrambled before Obama officially occupied the White House to enact an array of regulations aimed at weakening protections for workers and consumers, the environment, civil liberties, and abortion rights.[17] More than 90 rules were in play and most were designed to keep Bush's corporate friends happy by softening environmental protections such as the Clean Air Act, enabling a policy that "would make it much harder for the government to regulate toxic substances and hazardous chemicals to which workers are exposed on the job,"[18] and restoring "tax breaks for banks that take big losses on bad loans inherited through acquisitions."[19] As Jesse Jackson rightly points out, the fading Bush ideologues before the end of their tenure attempted to "make changes that the new administration will find it hard to reverse for years to come." Such changes would mean "more emissions from power plants; more exemptions from environmental-impact statements; [and] permission to operate natural gas lines at higher levels of pressure," among other "calamities."[20] And while an Obama administration may slow down the most egregious effects of 40 years of neoliberal rule over the economy and public life, neither a change in governance nor the existing near economic collapse guarantees a significant shift in the culture of neoliberalism and the influence of corporate power in maintaining a commanding influence over both the American economy and governmental politics. The market fundamentalism that "combines radical free-market ideology with the privatization of public wealth, the reduction of social welfare and the deregulation of economic activity" may be humbled as a result of the current economic crisis, but it is still alive and well as a major force "for transferring existing public and common wealth into private hands."[21] This is nowhere more evident than in the actions that allowed billions of public dollars to be handed out to Wall Street in 2009.

Notwithstanding the historic 2008 presidential election and the end of the Bush administration, neoliberal economics still represents a powerful force in American life, particularly in its beliefs that the common good should be coded as a pathology, market rationality is the ultimate embodiment of freedom, and, in spite of our current credit and financial crises, the future is best understood as a short-term investment defined by the bottom line. In spite of the threat of an economic meltdown, President Bush, just before leaving office, acted as if the regime of neoliberal capitalism was the only legitimate way to address the very crisis it had created. Shortly before the heads of the Group of 20 met in Washington near the end of 2008 for a summit on the global financial crisis, Bush gave a speech in which he once again affirmed his ideological loyalty to predatory capitalism and its illusory free-market wonders. First, he provided mindless praise for the very market fundamentalism that produced global market turmoil with his unapologetic boast, "I'm a market-oriented guy."[22] Second, seemingly clueless about the nature of the worldwide financial crisis, he proffered what sounded like a joke from Comedy Central's *The Daily Show*, hosted by Jon Stewart, insisting that "[t]he great threat to economic prosperity is not too little government involvement in the market. It is too much government involvement in the market."[23] In short, Bush repeated what has become an element of common sense in the public mind—that the market rather than politics gives people what they want. There is more at work here than the rallying cry of hard-line market fundamentalists, many of whom have been appointed to high-level cabinet positions in the Obama administration. There is also the presupposition that free-markets alone should provide for the welfare of human beings, suggesting not only an attack on the social state and a government willing to intervene on behalf of its citizens but also a notion of governing in which social needs are subordinated to economic interests. According to neoliberal economic policies, the welfare of human beings should be handed over to market forces, a presupposition that empties out the very constitutive nature of politics and the crucial realm of the social. But more is at risk here than the emptying out of politics: there are also the ravaging effects of a market fundamentalism that causes massive disparities in wealth and power along with the weakening of worker protections and the destruction of the social state, all of which are legitimated through a self-serving historical reinvention in which the success of politics is measured by the degree to which it evades any sense of social responsibility and public commitment. In this case, corporate sovereignty not only makes power invisible, it also conveniently erases a history of barbaric greed, unconscionable economic inequity, scandal-plagued politics, resurgent monopolies, and an unapologetic racism.[24]

What is often ignored by many theorists who analyze the rise of neoliberalism in the United States is that it is not only a system of economic power relations but also a political project, intent on producing new forms of subjectivity and sanctioning particular modes of conduct.[25] In addressing the absence of what can be termed the cultural politics and public pedagogy of neoliberalism, I want to begin with a theoretical insight provided by the British media theorist Nick Couldry, who insists that "every system of cruelty requires its own theater," one that draws upon the rituals of everyday life in order to legitimate its norms, values, institutions, and social practices.[26] Neoliberalism represents one such system of cruelty that is reproduced daily through a regime of common sense and a narrow notion of political rationality that "reaches from the soul of the citizen-subject to educational policy to practices of empire."[27] What is new about neoliberalism in the last three decades is that it has become normalized—even celebrated by the dominant media—and now serves as a powerful pedagogical force that shapes our lives, memories, and daily experiences, while attempting to erase everything critical and emancipatory about history, justice, solidarity, freedom, and the meaning of democracy. Undoubtedly, neoliberal norms, practices, and social relations are being called into question to a degree unwitnessed since the 1970s. Yet despite the devastating impact of a financial Katrina, neoliberalism's potent market-driven rationality is far from bankrupt and is still a powerful economic, cultural, and political force to be reckoned with.

Today, what we see spread out across this neoliberal landscape are desolate communities, gutted public services, weakened labor unions, 37 million impoverished people, 45 million Americans without health insurance, and a growing number of either unemployed or underemployed workers. If the Gilded Age returned with a vengeance in the first decade of the new millennium, so did an older legacy of rampant unregulated capitalism, merger mania, and a new class of Robber Barons dressed up as corporate power brokers with enormous political influence. Like its nineteenth-century counterpart, the New Gilded Age is marked by an obscene concentration of wealth among the privileged few while the number of poor Americans increases and inequality reaches historic high levels. The collapse of the markets in October 2008 promises only to contribute to this growing inequality between the rich and the poor.

## DISPOSABLE POPULATIONS

The varied populations devalued and made disposable under neoliberalism occupy a globalized space of ruthless politics in which the categories of "citizen" and "democratic representation," once integral to national politics, are no longer recognized. In the past, people who were marginalized by class and race could at least expect a modicum of support from the social state, either through an array of limited social provisions or from employers who recognized that they still had some value as part of a reserve army of unemployed labor. This is no longer true. Under the ruthless dynamics of neoliberal ideology, there has been a shift away from the possibility of getting ahead economically and living a life of dignity to the much more deadly task of struggling to stay alive. Many now argue that this new form of biopolitics—the politics that determine the life and death of human beings—is conditioned by a permanent state of class and racial exception in which, as Achille Mbembe asserts, "vast populations are subject to conditions of life conferring upon them the status of living dead."[28]

Disposable populations are increasingly relegated to the frontier zones and removed from public view. Such populations are often warehoused in schools that resemble boot camps,[29] dispersed to dank and dangerous workplaces far from the enclaves of the tourist industries, incarcerated in prisons that privilege punishment over rehabilitation, and consigned to the status of the permanently unemployed. Rendered redundant as a result of the collapse of the social state, a pervasive racism, a growing disparity in income and wealth, and a take-no-prisoners neoliberalism, an increasing number of individuals and groups are being demonized, criminalized, or simply abandoned, either by virtue of their status as immigrants or because they are young, poor, unemployed, disabled, homeless, or stuck in low-paying jobs. What Orlando Patterson in his discussion of slavery called "social death" has now become the fate of more and more people as the socially strangulating politics of hyperindividualism, self-interest, and consumerism become the organizing principles of everyday life.[30]

The harsh realities of this dehumanizing process is captured in a story told by Chip Ward, a thoughtful administrator at the Salt Lake City Public Library, who writes poignantly about his observations of a homeless woman named Ophelia. Ophelia spends time at the library because, like many of the homeless, she has nowhere else to go to use the bathroom, secure temporary relief from bad weather, or simply be able to rest. Excluded from the American dream and treated as both expendable and a threat, Ophelia, in spite of her mental illness, defines her own existence using a chilling metaphor. Ward describes Ophelia's presence and actions in the following way:

Ophelia sits by the fireplace and mumbles softly, smiling and gesturing at no one in particular. She gazes out the large window through the two pairs of glasses she wears, one windshield-sized pair over a smaller set perched precariously on her small nose. Perhaps four lenses help her see the invisible other she is addressing. When her "nobody there" conversation disturbs the reader seated beside her, Ophelia turns, chuckles at the woman's discomfort, and explains, *"Don't mind me, I'm dead. It's okay. I've been dead for some time now."* She pauses, then adds reassuringly, *"It's not so bad. You get used to it."* Not at all reassured, the woman gathers her belongings and moves quickly away. Ophelia shrugs. Verbal communication is tricky. She prefers telepathy, but that's hard to do since the rest of us, she informs me, "don't know the rules."[31]

Ophelia represents just one of the 200,000 chronically homeless who now use public libraries and other accessible but shrinking public spaces to find shelter. Many are often sick, addicted to drugs and alcohol, or mentally disabled, and many are close to a nervous breakdown because of the stress, insecurity, and danger they face daily. Along with the 1.6 million human beings who experience homelessness each year in the United States, they are treated like criminals, as if punishment is the appropriate civic response to poverty, mental illness, and human suffering. And while Ophelia's comments may be dismissed as the ramblings of a crazy woman, they point to something much deeper about the current state of American society and its abandonment of entire populations that are now considered the human waste of a neoliberal social order. Ward's understanding of Ophelia's plight as a public issue is instructive. He writes:

Ophelia is not so far off after all—in a sense she is dead and has been for some time. Hers is a kind of social death from shunning. She is neglected, avoided, ignored, denied, overlooked, feared, detested, pitied, and dismissed. She exists alone in a kind of social purgatory. She waits in the library, day after day, gazing at us through multiple lenses and mumbling to her invisible friends. She does not expect to be rescued or redeemed. She is, as she says, "used to it." She is our shame. What do you think about a culture that abandons suffering people and expects them to fend for themselves on the street, then criminalizes them for expressing the symptoms of illnesses they cannot control? We pay lip service to this tragedy—then look away fast.[32]

A more visible register of the politics of disposability at work in American society can be found in the haunting images of New Orleans following Hurricane Katrina: images of dead bodies floating in flooded streets and of thousands of African Americans marooned on highways, abandoned in the Louisiana Superdome, and waiting for days to be rescued from the roofs of flooded houses. Three years after Katrina, the politics of disposability returned to New Orleans with a vengeance and without apology as it was revealed that the Federal Emergency

Management Agency (FEMA) had received multiple warnings about dangerous levels of formaldehyde in the trailers they had provided for the victims of Hurricane Katrina, but the government agency refused to "conduct testing of occupied trailers because testing would 'imply FEMA's ownership of this issue.' "[33] Under the biopolitics of neoliberalism, conditions have been created in which moral responsibility disappears and politics no longer advocates for compassion, social justice, or the fundamental provisions necessary for a decent life. In this scenario, freedom is transformed into its opposite for most of the population as a small, privileged minority can purchase time, goods, services, and security while the vast majority is increasingly relegated to a life without protections, benefits, and support. For those populations considered expendable, redundant, and invisible by virtue of their race, class, and youth, life becomes increasingly precarious.

The collateral damage that reveals the consequences of this narrative of punishment and disposability becomes clear in heartbreaking stories about young people who literally die because they lack health insurance and live in extreme poverty. In one case, Deamonte Driver, a seventh grader in Prince George's County, Maryland, died because his mother did not have the health insurance to cover an $80 tooth extraction. Because of a lack of insurance, his mother was unable to find an oral surgeon willing to treat her son. By the time he was admitted and diagnosed in a hospital emergency room, the bacteria from the abscessed tooth had spread to his brain and, in spite of the level of high-quality intensive treatment he finally received, he eventually died. As Jean Comaroff points out in a different context, "the prevention of . . . pain and death . . . seems insufficient an incentive" to advocates of neoliberal market fundamentalism "in a world in which some 'children are . . . consigned to the coffins of history.' "[34]

## THE PLIGHT OF YOUTH

The weakening of the social state due to an onslaught of antidemocratic tendencies raises fundamental questions about not only the health of democracy in America but also what it might mean to take the social contract seriously as a political and moral referent in order to define the obligations of adults and educators to future generations of young people. For over a century, Americans have embraced as a defining feature of politics the idea that all levels of government would assume a large measure of responsibility for providing the resources, social provisions, and modes of education that would enable young people to prepare in the present for a better future, while expanding the meaning and depth of an inclusive democracy.[35] This was particularly true under the set of policies inaugurated under President Lyndon Johnson's Great Society programs of the 1960s, which were designed to eliminate both poverty and racial injustice.

Taking the social contract seriously, American society exhibited at least a willingness to fight for the rights of children, enact reforms that invested in their future, and provide the educational conditions necessary for them to be critical citizens. Democracy was linked to the well-being of youth, while how a society imagined democracy and its future was contingent on how it viewed its responsibility toward future generations. But as the United States, particularly under the Bush regime, became increasingly more authoritarian in its role as a national (in)security state, its use of surveillance, its suspension of civil liberties, its plundering of public goods, its suspension of basic social services, and its increasing use of torture and pure thuggery on the political level, it became clear that the current generation of young people was no longer viewed as an important social investment or as a marker for the state of democracy and the moral life of the nation.

Young people have become a generation of suspects in a society destroyed by the merging of market fundamentalism, consumerism, and militarism. Instead of a federal budget that addresses the needs of children, the United States has enacted federal policies that weaken government social programs, provide tax cuts for millionaires and corporations, and undercut or eliminate basic social provisions for children at risk. As *New York Times* op-ed columnist Paul Krugman points out, compassion and responsibility under the Bush administration gave way to "a relentless mean-spiritedness" and to the image of "President Bush as someone who takes food from the mouths of babes and gives the proceeds to his millionaire friends." For Krugman, Bush's budgets resembled a form of "top-down class warfare."[36] The dire consequences of the war against youth were apparent not only in the cutting of programs that benefit young people but also in President Bush's willingness to veto the State Children's Health Insurance Program in 2007, which provided much-needed health insurance to low-income children who do not qualify for Medicaid. As a result of this veto, "nearly one million American children . . . lose their health insurance."[37] And without any irony intended, Bush attempted to legitimate this disgraceful action by claiming that the bill would have "open[ed] up an avenue for people to switch from private insurance to the government."[38] Bush's actions gave new meaning to the neoliberal mantra "privatize or perish."[39]

Viewing the more recent U.S. federal government budget for 2009 as a political and ideological statement, it becomes clear that children continue to constitute one of the nation's lowest priorities. Marian Wright Edelman, president of the Children's Defense Fund, argues that the 2009 budget represents more than an act of unethical neglect, it constitutes an actual assault on children who are poor and impoverished. For example, while over 13 million children live in poverty, the budget threatens to "increase the number of children in America who are poor, uninsured, and lack access to quality early childhood education programs."[40] The 2009 budget cuts funds for Medicare, children's health insurance programs, the Emergency Medical Services for Children program, Head Start, vital nutrition programs, and housing vouchers for the poor. In addition, it proposes to eliminate "food stamps for more than 300,000 people in low-income families with children [and] funding for the Commodity Supplemental Food Program that would halt the distribution in an average month of nutritious food packages to more than 473,000 low-income mothers, children under age six, and seniors."[41] Fortunately, President Obama in his first 100 days in office has put into law an economic stimulus program that not only reverses many of these cuts but provides additional funding for the poor, unemployed, elderly, and other disadvantaged groups. Yet these reforms do little to address the toxic mix of poverty, homelessness, lack of health care for millions, the expanding ranks of the current 6 million people unemployed, and the deteriorating quality of public schools, especially for poor and minority children.

Insofar as these federal policies encode the prioritization of markets over people on a national scale, it is clear that political culture rejects any ethical commitment to provide young people with the prospects for a decent and just future. As expected, the current financial crisis will exacerbate even more the hardships that youth will experience in the future. In spite of the optimism accompanying the election of Barack Obama to the presidency of the United States, youth in America increasingly constitute a series of disappearances, badly represented in the public domain and largely invisible in terms of their own needs and as a reminder of adult responsibility. All young people today are increasingly defined, if not assaulted, by market forces that commodify almost every aspect of their lives and lived relations, though different groups of young people bear unequally the burden of a ruthless neoliberal order. Those young people on the margins of power who are viewed as flawed consumers or who resist the seductions of the commodity market increasingly fall prey to the dictates of a youth punishment-and-control complex that manages every aspect of their lives and increasingly governs their behavior through the modalities of surveillance and criminalization.

The dominant media now habitually reinforce the public perception of young people as variously lazy, stupid, self-indulgent, volatile, dangerous, and manipulative. These representations are present in television dramas such as *Gossip Girl*, in newspaper stories about youth violence, and on news programs that appear to take delight in showcasing the seamy side of youth culture. The American public is relentlessly treated to stories about how American children don't have a grasp of basic modes of history, language, and mathematics, and yet there is a deafening silence in most of the reports about how conservative policies have systematically disinvested in public schools, turning them largely into dull testing centers for middle-class students and warehousing units and surveillance centers for working-class and poor youth of color.[42] Even the highly regarded television program *60 Minutes* produced a 2006 episode about youth that suggested this kind of demonization. Highlighting the ways in which young people alleviate their alleged boredom, the show focused on the activity of "bum hunting," in which young people search out, attack, and savagely beat homeless people while videotaping the event in homage to the triumph of reality television. These acts are clearly reprehensible; but it is also reprehensible to vilify young people by suggesting that such behavior is in some way characteristic of youth in general. Then again, in a society in which politicians and the marketplace limit the roles available for youth to those of consumer, object, or billboard to sell sexuality, beauty products, music, athletic gear, clothes, and a host of other products, it is not surprising that young people are so easily misrepresented.

While all young people have to bear the consequences of a diminishing public concern about their care, dignity, and future, adult indifference and disrespect bear down on some youth much harder than on others. There is a long history in the United States of youth, particularly those of color, being associated in the media and by dominant politicians with a rising crime wave. What is really at stake in this discourse is the emergence of a punishment wave, one that reveals a society that does not know how to address those social problems that undercut any viable sense of agency, possibility, and future for many young people. We see antecedents of this assault in the work of John J. Dilulio, Jr., a former Bush adviser, who argued in an influential article published in the conservative *Weekly Standard* that American society faced a dire threat from an emerging generation of youth between the ages of 15 and 24, whom he called "super-predators."[43] Over the last decade, this misrepresentation of young people has been repeated in a series of Hollywood movies such as *Thirteen*, *Hard Candy*, *Alpha Dogs*, and *American Teen*, which routinely represent youth as dangerous, unstable, or simply without merit. When not portrayed as a threat to the social order, youth are often rendered as mindless, self-absorbed, and incapable of long-standing commitments. The *New York Times* ran two stories in 2008 about youth who exemplify this position. The first was a *New York Times* magazine cover story written by Emily Gould, whose narrative attempts to provide a public face to the mysterious world of the "professional" blogger.[44] Instead, the effort collapses into little more than a self-promoting blog post and an "orgy of shallow candor"[45] in which the author never once steps outside of the narrow world of her personal experiences—none of which even remotely gestures toward real social problems faced by her generation, problems that this faux freedom of the Internet is supposed to compensate for. Instead, we learn about the most indiscriminate elements of her personal life as filtered through virtual reality, intimate details about almost everyone she meets, and comments about celebrities, which she admits are mostly petty and cruel. In the end, her very public indiscretions and ethical violations prompt her to question whether making her "existence so public" could be taken back or erased. But, alas, not really, because we soon learn that for that to happen she would "have to destroy the entire Internet." Instead, she settles for shutting her eyes and praying "for an electromagnetic storm that would cancel out every mistake I'd ever made."[46] The second story, aptly titled "Let's Not Get to Know Each Other Better," is written by a young man who identifies himself as a "staunch proponent of a generation" that refuses to date, believes that "caring is creepy," and celebrates the "perfect hookup," which amounts to potential sexual encounters in which nothing is planned, and no commitments are allowed.[47] Moreover, we learn that "his generation" believes that "[t]he idea that two people can be happy together, maturing alongside each other, seems as false as a fairy tale." But, then again, why expect more from a generation of young people who he claims have "short attention spans [that] tend to be measured in nanoseconds ... [and who] float from room to room watching TV, surfing the Internet, playing Frisbee, and finding satisfaction around every corner, if only for a moment."[48] These two young people seem to embody perfectly a generation that is out of place, resembling insensitive and dangerous parasites feeding off the goodwill and largess of adult society. In other words, what both stories embody and reinforce is the growing public perception that aggression, abuse, contempt for the most basic social values, and a rabid individualism now characterize young people. In these scandalous representations of middle-class youth, kids are not armed and violent, they are simply stupid, clueless, cruel, and obsessed with using the public sphere to narrate themselves and play out their "emotional preoccupations and neurosis."[49]

While youth have been increasingly removed from the register of public concern, civic commitment, and ethical responsibility—viewed as a bad social investment—they linger in the public imagination as dim-witted, if not dangerous, ingrates, unworthy of compassion and so justifiably relegated to the civic rubbish pile. For example, a 2005 Associated Press - Ipsos poll found that "nearly 70 percent of Americans said they believed that people are ruder now than they were 20 or 30 years ago, and that children are among the worst offenders."[50] A similar disdain for children was visible in a 2008 trial in Brooklyn, New York, in which a father was convicted of murdering his seven-year-old daughter, who was described in the press as battered, starved, and weighing only "36 pounds—the same as a healthy child half her age," for eating a container of yogurt without his permission.[51] The father's defense lawyer argued that the seven year-old needed to be corrected, hence suggesting that she was responsible for her own death. According to the defense, she had the audacity to refuse to "be disciplined, slipping the ropes that bound her to the chair in her room, just out of reach of the litter box she was forced to use as a toilet. 'She was a little Houdini,' said the lawyer, Jeffrey T. Schwartz."[52] It is impossible to imagine this argument being made, or for even a moment entertained, in a court of law in a society that takes its responsibility to young people seriously.

The cumulative results of these narratives, images, and representations speak to a society in which the importance of both social bonds and learning how to make critical judgments and assume responsibility is easily forgotten, making it possible to view young people as a generation of suspects. What is also forgotten, as Susan Searls Giroux points out, is that "whatever undesirable features we assign to [youth] are more precisely a function of the world they have inherited, as shaped by adult decision—a world marred by extreme uncertainty, instability, volatility, and war."[53] Youth now represent the greatest affront to adult society because they have become the ultimate figure of the unspeakable—with the equally unthinkable catastrophic consequence of becoming disposable in a neoliberal society and a militarized state in which instrumental reason, finance capital, market rationality, instant gratification, deregulation, and a contempt for all things public, including public values, have reigned supreme for the last thirty years. What is evaded in these representations of young people is the recognition that the lives, experiences, and environments of the current generation of youth are entirely different from those of previous generations and that underlying these differences are various political, cultural, and social forces in which young people are considered unworthy of care, targeted, and relegated to a biopolitics of neoliberalism that privatizes reason and exhibits a disdain for all collective undertakings, especially those that address social responsibility and solidarity. Within this narrow individualism in which all that matters is one's ability to compete and "win" as defined by the ideologies, values, materials, social relations, and practices of commerce, it becomes difficult for young people to imagine a future in which the self becomes more than a self-promoting commodity and a symbol of commodification.

Unfortunately, such representations do more than degrade young people and resonate with their underlying marginality and disposability, they also legitimate the passage of draconian measures, policies, and laws at the highest levels of government. Thus it becomes increasingly difficult for young people to cope with the deep and massive downward shifts in the U.S. economy; in such an inhospitable climate, they are offered little help in the form of government policies to enable them to afford health care, college tuition, rising rents, or escalating mortgages. Even under the Obama administration, whatever increase in financial help young people receive for paying off college loans or relief from student debts, it is not enough. Anya Kamenetz, commenting on the plight of a generation deeply in debt, captures the profound sense of injustice and despair felt by youth who now face conditions unimaginable to previous generations. She writes:

I am 24 years old, and I was born into a broke generation. I look around and I see people who have borrowed more than they can repay to go to college, who can't find a good job, can't save, can't afford basic necessities like health insurance, can't make solid plans. Their credit card bills mount every month, while their lives stall out on the first uphill slope. Born into a century of unimaginable prosperity, in the richest country in the world, those of us between the ages of 18 and 35 have somehow been cheated out of our inheritance.[54]

Within a climate of economic and educational uncertainty, black youth are especially disadvantaged. Not only do a mere 42 percent who enter high school actually graduate, but they are increasingly jobless and marked as a surplus and disposable population in an economy that does not need their labor. Bob Herbert argues that "black American males inhabit a universe in which joblessness is frequently the norm [and that] over the past few years, the percentage of black male high school graduates in their 20s who were jobless has ranged from well over a third to roughly 50 percent.... For dropouts, the rates of joblessness are staggering. For black males who left high school without a diploma, the real jobless rate at various times over the past few years has ranged from 59 percent to a breathtaking 72 percent." He further argues, "These are the kinds of statistics you get during a depression."[55] Youth marginalized by class and color can no longer inhabit public spheres that allow them to take refuge behind their status as developing children worthy of adult protection and compassion. Whether it be the school, the community center, the street corner, or their place of residence, the most powerful and influential forces shaping their lives emanate from the security state and the criminal justice system. Increasingly there are more police in their schools than teachers, more surveillance cameras in their neighborhoods than public spaces that afford privacy and meaningful social interactions, and more liquor stores than health care centers, community outreach facilities, and recreational centers combined. The racialized spaces of oppression that poor youth of color inhabit make a mockery of the much-vaunted claim that the election of Barack Obama to the presidency suggests that institutionalized racism is over. In a neoliberal political order, with its celebration of radical individualism, privatization, and deregulation, any invocation of race can only be affirmed as a private prejudice, decoupled from wider institutional forces. This depoliticizing and privatizing of racism makes it all the more difficult to both identify the racialized attacks on poor youth of color and take the kind of action that would dismantle the systemic conditions that promote such practices of exclusion and disposability.

Punishment and fear have replaced compassion and social responsibility as the most important modalities mediating the relationship of youth to the larger social order. Youth within the last two decades have come to be seen as a source of trouble rather than as a resource for investing in the future, and in the case of poor black and Hispanic youth are increasingly treated as either a disposable population, cannon fodder for barbaric wars abroad, or the source of most of society's problems. Hence, young people now constitute a crisis that has less to do with improving the future than with denying it. As Larry Grossberg points out, "It has become common to think of kids as a threat to the existing social order and for kids to be blamed for the problems they experience. We slide from kids in trouble, kids have problems, and kids are threatened, to kids as trouble, kids as problems, and kids as threatening."[56] This was exemplified when the columnist Bob Herbert reported in the *New York Times* that "parts of New York City are like a police state for young men, women, and children who happen to be black or Hispanic. They are routinely stopped, searched, harassed, intimidated, humiliated and, in many cases, arrested for no good reason."[57] No longer "viewed as a privileged sign and embodiment of the future,"[58] youth are now increasingly demonized by the popular media and derided by politicians looking for quick-fix solutions to crime and other social ills. While youth have always had to bear the misplaced fear and distrust of adults, how youth are represented, talked about, and treated has changed dramatically in the last two decades.

Under the reign of neoliberal politics with its hyped-up social Darwinism and theater of cruelty, the popular demonization and "dangerousation" of the young now justifies responses to youth that were unthinkable 20 years ago, including criminalization and imprisonment, the prescription of psychotropic drugs, psychiatric confinement, and zero tolerance policies that model schools after prisons. School has become a model for a punishing society in which children who commit a rule violation as minor as a dress code infraction or slightly act out in class can be handcuffed, booked, and put in a jail cell. Racism, inequality, and poverty are on full display in the growing resegregation of public schools in the United States. Now more than ever, many schools either simply warehouse young black males or put them on the fast track to prison incarceration or a future of control under the criminal justice system. All across America, black and brown youth are being suspended or expelled at rates much higher than their white counterparts who commit similar behavioral infractions. For example, as Howard Witt writes in the *Chicago Tribune*, "In the average New Jersey public school, African-American students are almost 60 times as likely as white students to be expelled for serious disciplinary infractions. In Minnesota, black students are suspended 6 times as often as whites [and ] in Iowa, blacks make up just 5 percent of the statewide public school enrollment but account for 22 percent of the students who get suspended.... And on average across the nation, black students are suspended and expelled at nearly three times the rate of white students."[59] As schools become increasingly militarized, drug-sniffing dogs, metal detectors, and cameras have become common features in schools, and administrators appear more willing if not eager "to criminalize many school infractions, saddling tens of thousands of students with misdemeanor criminal records for offenses such as swearing[,] disrupting class," or pushing another student. Trust and respect now give way to fear, disdain, and suspicion, creating an environment in which critical pedagogical practices wither, while pedagogies of surveillance and testing flourish.[60] If young people were once defined as part of the vocabulary of innocence and compassion, they are now largely understood through the discourse of fear, guilt, and punishment.

Clearly, there is more at stake under the current regime of neoliberal biopolitics than an attack on children largely characterized by "negative labels and characterizations of youth [that] are falsely totalizing"[61] and punitive laws and public policies. Youth have also become collateral damage for conservatives and neoliberal advocates who want to dismantle the social state and in doing so justify themselves by pointing to an alleged rise of a generation of disorderly and dangerous youth dependent upon government entitlements. Within this discourse, government support for young people is both undermined and inappropriately blamed for creating a generation of kids labeled as psychologically damaged, narcissistic, violent, and out of control. Scapegoating youth as both a generation of suspects and a threat to the social order allows conservatives and neoliberals to further privatize those public spheres that youth need, such as education and health care, while developing policies that move away from social investment to matters of punishment and containment. In this instance, the punishing state combines with the logic of the market to produce priorities and policies that disinvest in the future of children and assert a ruthlessness that largely treats them as reified commodities or disposable populations.[62] Both childhood and the state are now being reimagined in ways that reveal the priorities of a society that has fully embraced the reckless abandon of casino capitalism, where the only rules that matter are made to order by powerful corporations and rich investors. How else to interpret neoliberal-inspired government programs that in the midst of deepening inequality, rising levels of poverty, catastrophic increases in failed mortgages, and growing unemployment invest more in prisons than in public and higher education? Where does justice enter into policies in which the government is willing to spend billions to bail out mortgage lenders Freddie Mac and Fannie Mae but refuses to provide adequate funding to raise 13 million American children above federal poverty levels? What relationship to democracy does a government have when it gives huge tax breaks to corporations that sell junk food and sugar-filled drinks to kids in public schools while providing meager funds to address the unprecedented health problems now facing many young people in the United States?[63]

At the current time, solutions involving social problems have become difficult to imagine, let alone address. For many young people and adults today, the private sphere has become the only space in which to imagine any sense of hope, pleasure, or possibility. Culture as an activity in which young people actually produce the conditions of their own agency through dialogue, community participation, public stories, and political struggle is being eroded. In its place, we are increasingly surrounded by a "climate of cultural and linguistic privatization" in which culture becomes something you consume, and the only kind of speech that is acceptable is that of the fast-paced shopper.[64] In spite of neoconservative and neoliberal claims that economic growth will cure social ills, the language of the market has no way of dealing with poverty, social inequality, or civil rights issues. It has no respect for noncommodified values and no vocabulary for recognizing and addressing social justice, compassion, decency, ethics, or, for that matter, its own antidemocratic forms of power. It has no way of understanding that the revolutionary idea of democracy, as Bill Moyers points out, is not just about the freedom to shop, formal elections, or the two-party system, "but the means of dignifying people so they become fully free to claim their moral and political agency."[65] These are political and educational issues, not merely economic concerns.

It is more necessary than ever to register youth as a theoretical, moral, and political center of concern, even as it is increasingly evident that youth are one of our lowest national priorities. It is crucial to connect the current crisis in democracy to the war against young people. Doing so will remind adults of their ethical and political responsibility to invest in youth as a symbol for not only securing a democratic future but also keeping alive those elements of civic imagination, culture, and education that subordinate economic principles to democratic values. The category of youth may be one of the most important referents for beginning a critical examination about the pernicious consequences of a society driven by market values, one that not only abstracts young people from the future but shapes the present in a theater of war in which youth become the most innocent victims. Youth provide a powerful touchstone for a critical discussion about the long-term consequences of neoliberal policies, which undermine any viable notion of justice, equality, and freedom, while also gesturing toward those conditions that make a democratic future possible. Many young people are part of social movements that not only address these crucial issues but also provide a politics, modes of resistance, and connective relations that adults should take seriously as part of their own civic and political formation at the beginning of the new millennium.[66]

## FUTURE MATTERS

In opposition to the authoritarian politics that was given full expression during the former Bush administration and whose legacy provides a decisive challenge to President Barack Obama's administration, it is crucial to remember that the category of youth does more than affirm that modernity's social contract is rooted in a conception of the future in which adult commitment is articulated as a vital public service; it also affirms those vocabularies, values, and social relations that are central to a politics capable of defending vital institutions as a public good and nurturing a flourishing democracy. At stake here is the recognition that children and youth constitute a powerful referent for addressing war, poverty, education, and a host of other important social issues. Moreover, as a symbol of the future, youth provide an important moral compass to assess what Jacques Derrida calls the promises of a "democracy to come."[67] In recent years, young people have all too often become the "vanishing point" of moral debate, considered irrelevant because they are allegedly too young or excluded from civic public discourse because they are viewed either at best as an important market for profits or at worst as reckless and dangerous.

Under neoliberalism, the abdication of the government's responsibility to protect public goods from private threats further reveals itself in the privatization of social problems and the vilification of those who fail to thrive in this vastly iniquitous social order. Too many youth within this degraded economic, political, and cultural geography occupy a "dead zone" in which the spectacle of commodification exists alongside the imposing threat of massive debt, bankruptcy, the prison-industrial complex, and the elimination of basic civil liberties. Indeed, we have an entire generation of unskilled and displaced youth who have been expelled from shrinking markets, blue-collar jobs, and the limited political power granted to the middle-class consumer. Rather than investing in the public good and solving social problems, the state now punishes those who are caught in the downward spiral of its economic policies. Punishment, incarceration, and surveillance represent the new face of governance. Consequently, the implied contract between the state and citizens is broken, and social guarantees for youth, as well as civic obligations to the future, vanish from the agenda of public concern. As market values supplant civic values, it becomes increasingly difficult "to translate private worries into public issues and, conversely, to discern public issues in private troubles."[68] Within this utterly privatizing market discourse, alcoholism, homelessness, poverty, joblessness, and illiteracy are not viewed as social issues, but rather as individual problems—that is, such problems are viewed as the result of a character flaw or a personal failing, and in too many cases such problems are criminalized.

In order to strengthen the public sphere, we must use its most widespread institutions, undo their metamorphoses into means of surveillance, commodification, and control, and reclaim them as democratic spaces. Schools, colleges, and universities come to mind, because of their contradictions and their democratic potential, their reality and their promise, although they are not the only sites of potential resistance. In this book, I argue that youth as a political and moral category is central for engaging and reclaiming the purpose and meaning of education as a democratic public sphere. This means recognizing that a future in which democratic possibilities will flourish can become a reality only if young people are provided with the knowledge, capacities, and skills they need to function as social agents, active citizens, empowered workers, and critical thinkers. Such a task must begin by examining the degree to which antidemocratic tendencies now threaten the capacity of public schools and higher education to educate subjects who can think, act, and struggle for a future that does not repeat the authoritarian present. And such a task must include a recognition that while the issues currently facing American youth arise from a neoliberal biopolitics whose increasing sway may be most visible in the way the Bush regime and other social institutions enshrined Gilded Age reasoning in their policies toward young people, the effects of neoliberalism are in no way limited to a past era in U.S. history. The issue of democracy as a global movement is crucial for youth, just as the current issues facing youth must be included in any conceptualization of global democracy. No rigorous attempt to understand the plight of young people today can ignore the particular, local effects of the biopolitics of consumption and disposability, nor should it ignore the state of democracy and society on a global scale. This book, by addressing the precarious lives of young people in the United States, contributes an analysis of consumer culture, criminalization, and education to what must become a massive collaborative endeavor to halt the relentless expansion of neoliberalism across the globe, if global democracy is to remain a possibility.

In the first chapter of this book, I analyze how market forces and the pedagogies of consumption that circulate therein increasingly bear down on young people so as to both commodify them and undercut their possibilities for critical agency. Chapter 1 examines how market sovereignty produces an analytic of youth that is entirely at odds with the requisites for producing democratic modes of citizenship. And while it does not distinguish how market forces impact young people in different ways, it provides a broad theoretical and political template for understanding how neoliberal rationality produces a mode of ideology, values, and social relations that threaten the possibility for all young people to step outside of its normalizing ideology and imagine a more democratic and just global order.

While commodification represents one form of social death for all young people, those youth who are marginalized by virtue of their race and class bear the burdens of not only the narrow impositions of a market-driven commodified culture but also the harsh experiences of impoverishment and suffering that mark them as disposable and redundant populations. Chapter 2 shifts the register beyond the logic of commodification and posits how the sovereignty of the market impacts differently on those poor youth of color who are excluded from a cheerful participation in its diverse pleasures and seductions and who as a result are defined through the registers of disposability and social death. This chapter makes visible the harshest elements of the punishing state and the egregious policies it enacts in a number of sites to render poor youth both disposable and politically powerless. Consequently, chapter 2 explores the logic of disposability as the underside of commodification, the fate of those considered flawed consumers, unworthy of social protections because they are considered a liability and utterly disposable in a market-driven world. If the soft side of neoliberal politics punishes all youth by treating them largely as commodities, it reveals a much darker side by subjecting poor youth and youth of color to the harshest elements, values, and dictates of neoliberal ideology. White wealthy kids may labor under the narrow dictates of a commodity culture, but they are not incarcerated in record numbers, placed in schools that merely serve to warehouse the refuse of global capitalism, or subjected to a life of misery and impoverishment. Actually, they benefit in the long run, under a market-driven society, from the transfer of public funds into private hands. But for those disposable populations of young people who are poor, especially black and Latino youth, neoliberal politics governs them through an analytic of punishment, surveillance, and control. In this instance, the corporate state becomes the punishing state and certain segments of the youth population become objects of modes of governance based on the crudest forms of disciplinary control.

The third chapter considers the role that academics and institutions of higher education may take in addressing the crisis of youth and its relationship to politics and critical education. This chapter analyzes the multifaceted attacks various conservative groups are waging against youth by undermining academic freedom and the conditions that make critical teaching and learning possible. At issue here is how the role of the university might be defined as a democratic, if not defiant, public sphere even as it is likewise subjected both to a ruthless corporate logic that confuses training and patriotic correctness with education and to a right-wing attack on any vestige of critical thought.

In the fourth and final chapter, I provide a broader theoretical analysis of what I call the biopolitics of neoliberalism, examining it as not merely an economic discourse but also as an educational, cultural, and political discourse that has gutted the notion of the social state and produced a set of policies that lay the groundwork for a politics of disposability that has dire consequences for society at large, and especially for young people. Only by understanding the pervasive and all-embracing reach of neoliberalism and its new mode of politics does it become possible to grasp the contours of a new historical period in which a war is being waged against youth, one that offers no apologies because it is too arrogant to imagine any resistance. Conversely, only by drawing attention to the particular effects of neoliberalism on the lives of young people—the focus of the first three chapters of this book—can a face be given to the ravages of a morally bankrupt and pernicious doctrine that is spreading like a pestilence and infecting democracy in the United States and around the globe. Most importantly, this book, in general, points to a gap in the various theories, discourses, and critiques trying to comprehend the impact of the current financial and economic crisis upon young people, labor, and others marginalized because they are poor, old, sick, brown, black, or simply left on their own to deal with the savagery of the free-market fall-out. While there is much talk among progressives about inequality generated by economic institutions, finance capital, and the legacy of historical imbalances in resources, power, and wealth, there is very little talk about creating the conditions for individual and collective agency as a fundamental basis for building social movements. That is, we must imagine the ways and means that make it possible for people to believe that their participation in political life matters, that they have voices that count, that they can make history. The task of a reinvigorated left is in large part to foreground consistently and imaginatively the question of justice in ways that translate private issues into public concerns, break open common sense in the interests of critical and reflective sense, and struggle to bring into being the conditions that enable people to use their power responsibly to control and shape the basic forces that bear down on their lives. This is not merely a theoretical issue: this is a preeminently educational issue that is at the heart of any viable notion of politics and central to addressing the related crises of youth and democracy.

CHAPTER 1

# BORN TO CONSUME: YOUTH AND THE PEDAGOGY OF COMMODIFICATION

*The market economy, source of all our freedoms, focus of all of our hopes, repository of our faith in progress, now threatens to crush us. It has annulled all alternatives to itself, thereby destroying one of the most fundamental of the human needs it purports to answer—the freedom to change, to find other forms of social and economic organization, to discover fresh ways of answering need, to imagine another future, the better world which this world could have been.... The cost of the leashed and diminished freedoms of the market, its celebrated "freedom of choice", can now be seen as a consolation for our own incarceration.*

—Jeremy Seabrook, *Consuming Cultures: Globalization and Local Lives*[1]

Modern society's fascination with the culture of childhood has a long and complex history. This is reflected not only in the changing nature of its social formations and state institutions but also in its own self-understanding, as children constitute the primary index through which a society registers its own meaning, vision, and politics. As many theorists of youth have reminded us, one of the distinctive elements of modernity was its acknowledgment of and commitment to the ideal that "[a] civilized society is one which struggles to make the world better for its children."[2] Understood by the nineteenth century as innocent beings in need of socialization, learning, and protection, children became an important modern symbol of both collective responsibility for and obligation to a future that would ensure their well-being and development as productive and worthy members of society. In this tradition, however contradictory, youth assumed the status of an important social investment, a political referent for adult responsibility, and a moral measure of how a society self-consciously undertakes to shape a more democratic future. Especially crucial in this discourse was not just the iconic figure of the child as the concrete embodiment of the promise and hope of the coming democratic social state but also the symbolic imagery of youth as a "guarantee that the present has the power to shape (even if in unpredictable ways) that future."[3]

It is important to recall, however, that the ideals of the modern period were often belied by practice. This same era also saw extreme child labor abuses in addition to slavery, class inequities, and imperialism, which all had material effects on the lives of young people and which were at least eventually acknowledged as social evils. What we have lost in today's landscape of rampant free-market ideology is the inherently productive negotiations between democratic theory and practice that characterized modernity. Within the last thirty years, particularly with the rise of neoliberalism and its rigid economic orthodoxy, the future has been recalibrated according to market calculations, effectively dissolving its primary commitment to children and its democratic moorings. Modernity's seemingly unshakable faith in progress has given way to a culture of fear, as conditions of economic and social uncertainty have come to define the future less as a promise than as a looming threat to be devalued or expelled. One consequence of this disinvestment in the future is a dramatic shift in how American society views, talks about, represents, and engages young people. The complex machinery of pedagogy, media, and politics is now largely mobilized to demean and punish rather than protect and nurture children. For many young people the future is bleak; the roles now open to them, as defined by commodity markets, shift between slacker employees and flawed consumers, or simply fodder for the human waste-disposal industry.[4] The modernist legacy of investing in the health and well-being of children and their future in accordance with the social contract is now hooked up to a respirator gasping for breath. Collective supports and rights for young people are disappearing as the family, school, social state, and various civic institutions abdicate their former guardianship and no longer serve as the primary forces shaping children's lives. Instead, global corporations and the punishing state are now the dominant storytellers and influence in children's lives, shaping their futures according to the interests of the market. Not only has liberal modernity faltered on its promise to future generations, it has radically redefined the meaning of youth, the nature of social betterment, and democracy itself.

If youth once constituted a social investment in the future and symbolized the promise of a better world, they are now entering another stage in the construction of a global social order in which children are increasingly demonized and criminalized—subject to random strip searches and increased surveillance, forced into prostitution, sold into child slavery, abducted as child soldiers, and made victims of numerous other forms of violence. As objects of a low-intensity war without end waged by governments and global corporations, youth are now defined within the languages of criminalization and commodification,[5] their daily existence delineated within a permanent state of emergency, mediated by heightened economic exploitation, class inequality, and racial injustice. Such forces are not new, nor do they bear down on all youth in the same way. While all young people are impacted by neoliberalism and the punishing state, poor black and brown youth directly bear the burden of a society that criminalizes their behavior and undercuts the conditions that might enable them to live a life of dignity and hope. This issue warrants great concern and I take it up in detail in chapter 3.

Along with the radically diminished status of youth, what has shifted in the current historical moment is a refined and intensified new mode of sovereignty and politics. Moreover, as power is increasingly colonized by global corporate networks, the state not only relinquishes its traditional monopoly on sovereignty to the market but is transformed from a social state to a punishing state. When market sovereignty undermines state sovereignty, the relationship between power and politics is altered. As Zygmunt Bauman puts it, "The result . . . is the gradual separation between the power to act, which now drifts towards markets, and politics, which, though remaining the domain of the state, is progressively stripped of its freedom of manoeuver and authority to set the rules and be arbiter of the game. This is indeed the prime cause of the erosion of the state's sovereignty."[6] State sovereignty thus refashioned abdicates its obligations to democratic governance and wields its remaining powers in the interests of matters of national security, a culture of fear, and disciplinary functions designed to contain, order, and control its various populations, especially young people.

Just as American society moves into the twenty-first century, neoliberalism exercises its own form of sovereignty and mode of rationality through the invisible hand of the market, which is aided by a host of emerging electronic technologies and produces new modes of governance, new relationships between power and politics, and a distinct merging of everyday life and the realm of the political. The consequences of this new configuration of sovereignty and the power relations to which it gives rise are not only a crisis of modernity, as Lawrence Grossberg suggests, but a transformation in the very nature of politics.[7] Politics has become increasingly *biopolitical*, a shift crucial to understanding the changing conditions of youth in the United States and elsewhere around the globe. While the notion of biopolitics differs significantly among its most prominent theorists, including Michel Foucault, Giorgio Agamben, and Michael Hardt and Antonio Negri,[8] what these theorists share is a commitment to think through the convergence of life and politics, locating matters of "life and death within our ways of thinking about and imagining politics."[9] Biopolitics points to relations of power that are more capillary and capacious, concerned not only with the body as an object of disciplinary control but also with a body that needs to be "regularized," subject to those pedagogical modes of production that fashion whole ways of life and enlarge the targets of control and regulation.[10] Populations are now controlled not simply through the threat of force but through technologies of consent produced in a vast array of apparatuses extending from the school to the varied instances of screen and electronically mediated culture. Central to the new form of neoliberal biopolitics is the issue of how youth are to be defined, guided, constituted, governed, and at times abandoned in accordance with a market-based rationality and logic of accumulation. At stake here is both a mode of governmentality—a mode of power par excellence designed to produce a market-based notion of agency and subjectivity—and the emergence of a more intensified political economy organized by three principal concerns: deregulated markets, commodification, and disposability.[11] This is a politics that not only empties the concepts of citizenship,[12] the future, and democracy of any substantive content, but relegates entire populations to either the dystopia of consumerism or the dead zone of a "production line of human waste or wasted humans."[13] This shift in sovereignty and power now makes the mechanisms of biopolitical subject formation—those forces affecting matters of life, death, and survival—central to politics. But more than this, a new analytic of youth lies at the very heart of a biopolitics of neoliberal governance.

This new analytic of youth can be grasped in how the United States has been reconfigured in two ways. Within the last three decades, the social state has largely been replaced by the punishing state. But at the same time we have also witnessed the emergence of a mode of sovereignty in which the impersonal calculations of the market now wield the true levers of political, social, and economic power. Under this mode of market sovereignty, youth are not only subject to punishment, they are also relentlessly commodified. That is, they are the prime targets of yet another attack as they are largely redefined less through the kinder optic of democratic values and a democratic future than by the utterly reductive act of consuming central to a neoliberal society. One consequence is that youth in the new millennium are viewed neither as innocent and in need of protection nor as an important social investment for the future; instead, they have become part of what Bauman calls "an acute crisis of the human waste disposal industry,"[14] rendered unworthy of social and political rights and viewed as redundant and expendable.

In what follows, I will analyze how the current subjection of youth to omnipresent forces of commodification constitutes what Jacques Derrida calls an "autoimmunitary" logic, that is, a process whereby under a neoliberal biopolitics the United States has entered a period marked most notably by a war against young people. This war not only undermines the social contract but also erodes the democratic body politic by working in "quasi-suicidal fashion . . . to destroy its own protection, to immunize itself against its 'own' immunity."[15] Moreover, this autoimmunity process works to remove all those collective rights that provide young people with the promise of a just and desirable future; it also offers American society little protection against the antidemocratic tendencies of a market sovereignty predicated on dissolving all democratic modes of sociality, erasing all vestiges of the public good, turning citizens into consumers, and commodifying every aspect of the social order—while at the same time threatening global ecological sustainability. What is particularly troubling about this "figure of societal suicide"[16] is that it points to a society that is consuming itself by destroying its children as it simultaneously engages in a scale of planetary consumption that "is going on at a rate which literally cannot be sustained [and] threatens the very biological survival of humans and related species."[17]

## CONSUMERIST POLITICS AND PEDAGOGIES OF COMMODIFICATION

In the society of consumers no one can become a subject without first turning into a commodity, and no one can keep his or her subjectness secure without perpetually resuscitating, resurrecting and replenishing the capacities expected and required of a sellable commodity. The "subjectivity" of the "subject," and most of what that subjectivity enables the subject to achieve, is focused on an unending effort to itself become, and remain, a sellable commodity. The most prominent feature of the society of consumers—however carefully concealed and most thoroughly covered up—is the transformation of consumers into commodities; or rather their dissolution into the sea of commodities.[18]

—Zygmunt Bauman, *Consuming Life*

Lawrence Grossberg claims that "[e]conomics has become sexy, while politics has become vulgar [and that] the rise of economics [may be] due to the fall of politics."[19] In more specific terms, he insists that under the biopolitics of neoliberalism "[t]he free market... is fundamentally an argument against politics.... It is an argument against the power of the state to intervene on behalf of its citizens against the unpredictability or the ravages of the market."[20] Actually, if economics has become sexy, it is precisely due to a new kind of politics that privileges exchange value, resists all forms of government intervention (except when it benefits the rich and powerful or uses force to maintain social order), celebrates excessive individualism, and consolidates the power of the rich—the dreaded consequences of which are now becoming more visible as the current financial and economic crisis unfolds. But neoliberal politics is successful because it also works hard through the related modalities of education and seduction to produce a new kind of youthful biopolitical subject willing to conform to the narrow dictates, values, and dreams of totalizing market society. Under the neoliberal regime, an intense battle is being waged through a public pedagogy of consumerism designed to influence, shape, and produce future generations of young people who cannot separate their identities, values, and dreams from the world of commerce, brands, and commodities. Unlike the stripped-down and locked-down version of state sovereignty, this new biopolitics of market sovereignty makes matters of education, pedagogy, and the production of consent central to its mode of politics. Under market sovereignty, a culture of critical learning and engagement is replaced by a "culture of disengagement, discontinuity, and forgetting."[21] Whereas state sovereignty is largely focused on its policing, surveillance, and security functions, market sovereignty consolidates advertising and marketing practices; seduction and persuasion replace the panoptical model of power relations. What is unique and particularly disturbing about neoliberalism is that it makes undemocratic modes of education central to its politics and employs a mode of pedagogy aimed at displacing and shutting down all vestiges of the public sphere. To a greater extent than at any other point in liberal modernity, the regime of neoliberal biopolitics extends economic rationality now "to formerly noneconomic domains [shaping] individual conduct, or more precisely, [prescribing] the citizen-subject of the neoliberal order."[22] Most crucially, this struggle over the construction of the neoliberal consumer-subject, especially as it applies to young people, is by and large waged outside of formal educational institutions, in pedagogical sites and spaces that are generally privatized and extend from the traditional and new media to conservative-funded think tanks and private schools.[23] As commodity markets assume a commanding role "in raising, educating and shaping children,"[24] pedagogy is redefined as a tool of commerce aggressively promoting the commodification of young people. Free-market fundamentalism is more than willing to invest in a risky but repellent conceit that defines children's worth in largely market values, reducing them to both commodities and a source of profit. This suggests a system that is deeply immoral, rather than amoral, as some critics suggest.

Increasingly, moral and ethical considerations are decoupled from the calculating logic and consequences of all economic activity, and yet the commodification of young people is rarely challenged, suggesting that the biopolitics of neoliberalism and its model of consumerism have "become a kind of default philosophy for modern life."[25] In part, this is exemplified in not only the endless public pronouncements that make a market society and democracy synonymous, but also in the ongoing celebration, in spite of the economic meltdown, of the fundamental values of the new Gilded Age. The United States now struggles with an identity, aggressively promoted by the George W. Bush cabal, which celebrates its success in rolling back the New Deal and demolishing "big government." Although free-market euphoria may be tempered by the economic crisis of 2009, the collusion of multiple political and economic forces has worked toward the unfortunate return of an era when President Calvin Coolidge could state, without irony, to "a conference of admen that they were doing God's work," implying that the marketers and advertisers were at the forefront of the lines protecting the ethical foundations of American society, if not democracy itself.[26] Despite a change in U.S. political leadership, these forces—if left unchecked—will continue to carry on a transformation of democratic governance and citizenship until they are both completely destroyed.

The dismaying possibility that the ideals of a new Gilded Age have become permanently entrenched in the American social imaginary is suggested by one of the most popular television series in 2008, *Mad Men*, which follows the work rituals of a group of boozing, smoking, and misogynist marketing executives in the early 1960s—before the civil rights movement and student protests wrecked everything. Bathed in a nostalgic appeal for an era when endless consumption kept the country prosperous and advertising became a primary weapon in the Cold War as it allegedly demonstrated both the real meaning of freedom and capitalism's advantage over the communist threat,[27] the series endorses a less ruthless period of American consumerism, while carefully excluding the dark side of advertising's pernicious influence on children, democratic values, and public life. While the program does offer a critique of some of the ad executives, these criticisms are portrayed mostly as character flaws, while the ad industry in general is viewed as glamorous. More importantly, the popularity of *Mad Men* is symptomatic of a society that refuses to acknowledge that under the onslaught of a more pervasive and intensified consumer culture the material conditions of childhood in American life, if not democracy itself, have changed dramatically for the worse. If *Mad Men* offers up a nostalgic appeal to advertising's golden age, a reality television show, *The Apprentice*, hosted by the infamous CEO Donald Trump, provides viewers with a take-no-losers approach to how corporate culture works as a structure for organized violence. Presented as a contest to determine who is best qualified to run one of Trump's companies for a starting annual salary of $250,000, the program both introduces and sanctions duplicity, intolerance, and distrust as essential to competing successfully in the corporate world. Being hard, ruthless, and cruel is legitimated, with Trump as the embodiment of the ultimate celebrity CEO who entertains his audiences by encouraging job applicants to undercut each other and do whatever it takes to get the job, while punishing those who do not live up to his standards with a public humiliation before they are fired. In different ways, both programs display and legitimate, without apology, the increasingly discredited aggressiveness and antidemocratic ethos of corporate culture and management.

Even as this new market-driven society becomes devalued, what is still hidden under the mantle of neoliberal rationality camouflaged as common sense are the ways in which the most consumer-oriented society in the world is fundamentally altering the very experiences and hopes of young people, and often with tragic consequences. As Daniel Cook points out, "What is most troubling is that children's culture has become virtually indistinguishable from consumer culture over the course of the last century. The cultural marketplace is now a key arena for the formation of the sense of self and of peer relationships, so much so that parents often are stuck between giving into a kid's purchase demands or risking their child becoming an outcast on the playground."[28]

The underlying biopolitics of neoliberalism points to a war being waged against youth on yet another front. This is a front largely populated by ubiquitous screen cultures, electronic technologies, and landscapes of desire that mix image, text, and sound into new and powerful social formations and pedagogical practices. The animating project of neoliberalism is disturbingly focused on children and the creation of a consumer society that is less concerned with ensuring the future than with reconfiguring it through an economic logic in which "all dimensions of human life are cast in terms of a market rationality."[29] As the sites of entertainment, advertising, and education converge, youth occupy an entirely new place in the social order. Once proclaimed as innocent and in need of protection, they are now viewed as one of the central pillars of the consumer economy and increasingly are exposed to market concepts and relations in public spheres and areas of life that were once typically heralded as a safe haven from market values. While children have always been touched by market relations in capitalist societies, the terms, conditions, and reach of the biopolitics and practices of commodification are unlike anything we have seen in the past. As democracy is increasingly reduced to an empty shell and the carceral state looms heavy on the twenty-first-century horizon, the commodity form penetrates all aspects of daily life, shaping the very nature of how young people think, act, and desire, and marking them as the epicenter of consumer culture. And it is precisely this violence against children as part of an attempt to universalize the hyperindividual isolated subject of consumption that is one of the most neglected aspects of the study of the politics of neoliberalism, commodification, and disposability.

## CONSUMING AND THE POLITICS OF WASTE

A hundred years later it seems that a most fatal, possibly the most fatal, result of modernity's global triumph is the acute crisis of the human-waste disposal industry: with the volume of human waste outgrowing the extant managerial capacity, there is a plausible prospect of the new planetary modernity choking on its own waste products which it can neither reassimilate nor annihilate.[30]

—Zygmunt Bauman, *Wasted Lives*

What is distinct about the biopolitics of neoliberalism is that not only have corporate hierarchies, market values, and advertising practices

become a template for the entire society, but there has been a significant move away from a society marked by the primacy of production
to a society that largely " 'interpellates' its members first and foremost
in their role of consumers."[31] Zygmunt Bauman has brilliantly argued
in a number of books that with the emergence of consumerism as the
organizing principle of society, a social order has arisen in which not
only work has been superseded by the practices of consumption, but
the foundational values and principles of an older production-oriented
society have been replaced by a new set of market-driven priorities.[32]
In this consumer society, the modern political and economic ambi-
tions that stressed procrastination, delay, long-term investments, and
durability have been replaced by an emphasis on speed, instant gratification, fluidity, and disposability. Progress has become dystopian,
defined less by the rewards of social and economic mobility than
by the fear of being left behind in a global landscape marked by
"relentless and inescapable change [auguring] not peace and respite
but continuous crisis and strain, forbidding any moment of rest."[33]
Speed becomes crucial in a consumer society in which all goods, fashions, and trends are affirmed through the thrill of the quick turnover,
the immediacy of gratification, and the requisite short shelf life that
quickly qualifies them for the nearest garbage dump. Though writing
in a much different context, Jason Pine's literal and metaphoric comments on the way speed colonizes the body captures a deeply held
truth about consumer society. He writes: "[T]he high-speed loop of
consumption-production-consumption continues to approach a maximum velocity where the only thing that remains will be speed."[34]
Of course, there is more at stake here than ceaseless change; there is
also the suspension of judgment, the inability to think critically, the
avoidance of responsibility, the burst of pleasure that accompanies an
endless reserve of choices, and the liberation of choices from either
their consequences or public considerations.

For Bauman, the consumer society represents a shift from solid
modernity to "liquid modernity," a metaphor that points to "one
trait that all liquids share: the feebleness, weakness, brevity and frailty
of bonds and thus inability to keep shape for long."[35] Power and
authority are commanded less by the state than by a market that
entertains the public with proliferating commodities whose use-value
is measured by how quickly they can be discarded in order to make
way for new purchases. Long-term commitments along with values
related to durability are now sacrificed to a mode of temporality in
which quick turnovers and short attention spans become the measure
of how our everyday lives are experienced and futures anticipated.[36]

As authority is colonized by the market, politics loses its moral force to the appeal and status of the latest trend. Social identities are now carved out of brands, fashioned out of commodities, and used up as quickly as possible. Emotions remain shallow, quickly discharging as impulses, eschewing at all cost any purposeful thought and reflection. Utopia in a consumer society is utterly privatized and decidedly anticollectivist, defined less by the older conservative notion of freedom from interference than by the freedom to endlessly consume, a set of preferences in which commodification reigns supreme through the powerful influence and reach of the pedagogical machinery of the market.[37] Under "liquid modernity," consumption is no longer primarily about desiring or buying possessions but the pleasure of purging oneself of them. In this discourse, the related concepts of transience, circulation, replacement, waste, and disposability become central to a biopolitics of neoliberalism in which nothing holds its shape for long and everything is up for sale. According to Bauman:

Contrary to some scholarly accounts and popular beliefs, consumption (surely in its current form) is not about possessions, but about acquisition and quick disposal of the goods acquired so that the site is cleared for the next shopping expedition. Removal, dumping and destruction of goods are as crucial a part of consumption as the acquisition of goods. If production is about creation of "useful objects," objects with "use value," consumption is primarily about "using objects up," and so creation of waste, useless objects, sometimes toxic, often seen as dangerous and feared, but always viewed as degraded and degrading, a repellent and off-putting matter.... Both speed and fluidity are needed for consumerism to thrive. Fully-fledged consumers are not finicky about consigning things to waste—they accept the short life-span of things and their mortality with equanimity; the most seasoned among them learn even to rejoice in getting rid of things that have passed their use-by date. Consumer society cannot but be a society of excess and profligacy—and so of redundancy and prodigal waste.[38]

While the current economic and financial crisis has slowed the pace of consumption, it has done little to mount a robust challenge to the underlying values that produced it. Under the biopolitics of neoliberalism and its construction of the consuming life and society, politics removes itself from the public good while presenting an argument against any "politics, or at least against a politics that attempts to govern in social rather than economic terms."[39] The ideal of government as the servant of the people has been undermined by a corporate neoliberal ethos in which citizens have little to do with exercising power and shaping policy. Instead they often become both passive and complicitous in their own depoliticization, as consumer sovereignty empties democracy of any substance while offering citizen-consumers limited participation in the social order that prioritizes purchasing goods and making money over the demands of civic responsibility. The consumer society does more than produce a rationality and mode of agency in which people are ill-equipped to assume the responsibilities of caring for others and sustaining the promise of a democratic society; it also gives rise to what Bauman calls "wasted lives" and a politics of disposability.[40] As the sovereignty of the consumer replaces the sovereignty of the citizen, commodification extends its reach from products to human beings, from the act of production to all social relations. Under such circumstances, "[t]he society of consumers is unthinkable without a thriving waste-disposal industry."[41] But that waste encompasses not only the steady stream of objects, goods, and services for fast consumption and disposal; it now includes specific individuals and groups, many of whom fall into the category of failed consumers. For those marginalized others considered useless and superfluous, progress is no longer about getting ahead or improving the quality of their lives: it is about the struggle to survive and not to succumb to the status of the walking dead. With the social state in retreat, safety nets disappear just as more and more people lose their jobs, declare bankruptcy, become homeless, slip beneath the poverty line, lack health care, and understand their misfortunes as individual bad luck rather than as a socially induced problem. According to Bauman, these individuals and groups are the new waste products of "liquid modernity,"[42] those who have by default withdrawn or have been involuntarily removed from the rituals of the consumer society. They are the outlaws of the market, flawed as consumers and devalued as commodities, and they pose ethical and political problems for a neoliberal biopolitics that defines itself as the apogee of freedom and democracy. Unable to participate in the rituals of status-seeking consumption, those "othered" in a consumer society are subject to a "new lexicon of cultural domination and symbolic violence [that] distributes shame and humiliation to those lower down the hierarchy." They also bear "the pain of failure, of being a loser, of being invisible to those above, [all of which] cuts a deep wound in the psyche."[43] These new waste products of a consumer society include the poor, the jobless, immigrants, youth, and other individuals and groups who occupy a liminal space marked by insecurity, uncertainty, and deprivation. In Bauman's words,

They are truly and fully useless—redundant, supernumerary leftovers of a society reconstituting itself as a society of consumers; they have nothing to offer, either now or in the foreseeable future, to the consumer-oriented economy; they won't add to the pool of consumer wonders, they won't "lead the country out of depression," reaching for credit cards they don't have and emptying saving accounts they don't possess—and so the "community" would be so much better off were they to disappear.[44]

Against the glittering landscape of the second Gilded Age, waste disposal becomes difficult and more ominous as there are no social supports left to recycle those now considered redundant and dispensable. While in some cases the growing armies of "human waste" disappear into a world that is as invisible as it is unthinkable, it is more often the case that they cannot vanish or be absorbed into the dark alleys and shadows of the glittering consumer marketplace quite fast enough for affluent consumers unnerved by their presence. As the task of disposability becomes more difficult for governments, solutions are sought in the prison system, the final outpost for disposable populations.[45] As Bauman puts it,

The immediate proximity of large and growing agglomerations of "wasted humans," likely to become durable or permanent, calls for stricter segregationist policies and extraordinary security measures, lest the "health of society," the "normal functioning" of the social system, be endangered. The notorious tasks of "tension management" and "pattern maintenance" that, according to Talcott Parsons, each system needs to perform in order to survive presently boil down almost entirely to the tight separation of "human waste" from the rest of society, its exemption from the legal framework in which the life pursuits of the rest of society are conducted, and its "neutralization." "Human waste" can no longer be removed to distant waste disposal sites and placed firmly out of bounds to "normal life." It needs therefore to be sealed off in tightly closed containers.[46]

In the hyped-up consumer society of neoliberalism, the prevailing ideology insists that all aspects of the social order bear the imprint of the calculating logic of the market and succumb to commodification. The dreamscapes that make up a society built on the promises of mass consumption translate deftly into ad copy, insistently promoting and normalizing a neoliberal order in which economic relations now provide the master plan for defining human beings, their relations with others, and the larger world. As tokens of economic exchange, commodities circulate in the wider society as "both objects and as semi-magical and symbolic representations."[47] Goods now have meanings attached to them out of which individuals, groups, and institutions develop identities, mediate relations, and experience pleasure. The complex nature of the politics of consumerism and the explosion of the commodity form into all aspects of social life have been analyzed with great care by George Lukacs, the Frankfurt School, and Guy Debord—and more recently by Fredric Jameson, Jean Baudrillard, and Naomi Klein, among others.[48] The legacy of this work has provided an important theoretical service in critically analyzing the connection between transnational corporations and the exploitation of labor, the emergence of a one-dimensional society, the triumph of a culture of signs, and the growth of an image-soaked hyperreality that obliterates the distinction between the real and the imaginary. At the same time, it has not addressed in sufficient analytic fashion how a consumer society under the biopolitical regime of neoliberalism has made young people the target of its massive and diverse strategies of consumption and commodification and what the latter implies for rethinking the very nature of agency, democracy, and politics itself.[49] It is to this issue that I will now turn.

## THE BUSINESS OF COMMODIFYING KIDS

The commodity has penetrated every aspect of people's lives all over the world in ways that have no historical precedent. The commodity—and capitalism in general—has insinuated itself into structures of feeling, into the most intimate spaces of people's lives. At the same time human beings are more connected than ever before and in ways we rarely acknowledge. I am thinking of a song performed by Sweet Honey in the Rock about the global assembly line, which links us in ways contingent on exploitative practices of production and consumption. In the Global North, we purchase the pain and exploitation of girls in the Global South, which we wear everyday on our bodies.[50]
  —Angela Davis, *Abolition Democracy: Beyond Empire, Prisons, and Torture*

In a society that measures its success and failure solely through the economic lens of the Gross National Product (GNP), it becomes difficult to define youth outside of market principles determined largely by criteria such as the rate of market growth and the accumulation of capital. The value and worth of young people in this discourse is largely determined through the bottom-line cost-benefit categories of income, expenses, assets, and liabilities. The GNP does not measure justice, integrity, courage, compassion, wisdom, and learning, among other values vital to the interests and health of a democratic society. Nor does it address the importance of civic participation, public goods, dissent, and the fostering of democratic institutions. In a society driven almost entirely by market mentalities, moralities, values, and ideals, consuming, selling, and branding become the primary mode through which to define agency and social relations—intimate and public—and to shape the sensibilities and inner lives of adults as well as how society defines and treats its children. In what follows, I want to argue that under the biopolitics of neoliberalism, American society has undergone a sea change in the daily lives of children that marks a major transition from a culture of innocence and social protection, however imperfect, to a culture of commodification. This is culture that does more than undermine the ideals of a secure and happy childhood; it also exhibits the bad faith of a society in which, for children, "there can be only one kind of value, market value; one kind of success, profit; one kind of existence, commodities; and one kind of social relationship, markets."[51] Children now inhabit a cultural landscape in which they can only recognize themselves in terms preferred by the market. As Benjamin Barber argues:

The [childhood] consumer at once both imbibes the world of products, goods, and things being impressed upon her and so conquers it, and yet is defined via brands, trademarks, and consumer identity by that world.... The dollars or Euros or yen with which she imagines she is mastering the world of material things turn her into a thing defined by the material—from self-defined person into market-defined brand; from autonomous public citizen to heteronomous private shopper. The boundary separating her from what she buys vanishes: she ceases to buy goods as instruments of other ends and instead becomes the goods she buys.[52]

Subject to an advertising and marketing industry that spends over $17 billion a year on shaping children's identities and desires,[53] American youth are commercially carpet bombed through a never-ending proliferation of market strategies that colonize their consciousness and daily lives. Multi-billion-dollar corporations, with the commanding role of commodity markets as well as the support of the highest reaches of government, now become the primary educational and cultural force in shaping, if not hijacking, how young people define their interests, values, and relations to others. Juliet Schor, one of the most insightful and critical theorists of the commodification of children, argues that "[t]hese corporations not only have enormous economic power, but their political influence has never been greater. They have funneled unprecedented sums of money to political parties and officials.... The power wielded by these corporations is evident in many ways, from their ability to eliminate competitors to their ability to mobilize state power in their interest."[54]

As the sovereignty of the market displaces state sovereignty, children are no longer viewed as an important social investment or as a central marker for the moral life of the nation. Instead, childhood ideals linked to the protection and well-being of youth are transformed—decoupled from the "call to conscience [and] civic engagement"[55] and redefined through what amounts to a culture of cruelty, abandonment, and disposability. Childhood ideals increasingly give way to a market-driven politics in which young people are prepared for a life of objectification while simultaneously drained of any viable sense of moral and political agency. Moreover, as the economy implodes, the financial sector is racked by corruption and usury, the housing and mortgage market is in free fall, and millions of people lose their jobs, the targeting of children for profits takes on even more insistent and ominous tones. This is especially true in a consumer society in which children more than ever mediate their identities and relations to others through the consumption of goods and images. No longer imagined within language of responsibility and justice, childhood begins with what might be called the scandalous philosophy of money—that is, a logic in which everything, including the worth of young people, is measured through the potentially barbaric calculations of finance, exchange value, and profitability.[56]

What is distinctive about this period in history is that the United States has become the most "consumer-oriented society in the world." Kids and teens, because of their value as consumers and their ability to influence spending, are not only at "the epicenter of American consumer culture" but are also the major targets of those powerful marketing and financial forces that service big corporations and the corporate state.[57] In a world in which products far outnumber shoppers, youth have been unearthed not simply as another expansive and profitable market but as the primary source of redemption for the future of capitalism. Erased as future citizens of a democracy, kids are now constructed as consuming and saleable objects. Gilded Age corporations and their army of marketers, psychologists, and advertising executives now engage in what Susan Linn calls a "hostile takeover of childhood,"[58] poised to take advantage of the economic power wielded by kids and teens. With spending power increasing to match that of adults, the children's market has greatly expanded in the last few decades, in terms of both direct spending by kids and their influence on parental acquisitions. While figures on direct spending by kids differ, Benjamin Barber claims that "in 2000, there were 31 million American kids between twelve and nineteen already controlling $155 billion consumer dollars. Just four years later, there were 33.5 million kids controlling $169 billion, or roughly $91 per week per kid."[59] Schor argues that "children age four to twelve made . . . $30.0 billion" in purchases in 2002, while kids aged twelve to nineteen "accounted for $170 billion of personal spending."[60] Molnar and Boninger cite figures indicating that preteens and teenagers command "$200 billion in spending power."[61] Young people are attractive to corporations because they are big spenders, but that is not the only reason. They also exert a powerful influence on parental spending, offering up a market in which, according to Anap Shah, "Children (under 12) and teens influence parental purchases totaling over . . . $670 billion a year."[62]

One measure of the corporate assault on kids can be seen in the reach, acceleration, and effectiveness of a marketing and advertising juggernaut that attempts to turn kids into consumers and childhood into a saleable commodity. Every child, regardless of how young, is now a potential consumer ripe for being commodified and immersed in a commercial culture defined by brands. According to Lawrence Grossberg, children are introduced to the world of logos, advertising, and the "mattering maps" of consumerism long before they can speak: "capitalism targets kids as soon as they are old enough to watch commercials, even though they may not be old enough to distinguish programming from commercials or to recognize the effects of branding and product placement."[63] In fact, American children from birth to adulthood are exposed to a consumer blitz of advertising, marketing, educating, and entertaining that has no historical precedent. There is even a market for videos for toddlers as young as four months old. One such baby video called *Baby Gourmet* alleges to "provide a multi-sensory experience for children designed to introduce little ones to beautiful fruits and vegetables . . . in a gentle and amusing way that stimulates both the left and right hemispheres."[64] This would be humorous if Madison Avenue was not dead serious in its attempts to sell this type of hype—along with other baby videos such as *Baby Einstein, Brainy Baby, Sesame Street Baby*, and Disney's *Winnie the Pooh Baby*—to parents eager to provide their children with every conceivable advantage over the rest. Not surprisingly, this is part of a growing $4.8 billion market aimed at the youngest children.[65] Schor captures perfectly the omnipotence of this machinery of consumerism as it envelops the lives of very young children:

At age one, she's watching *Teletubbies* and eating the food of its "promo partners" Burger King and McDonald's. Kids can recognize logos by eighteen months, and before reaching their second birthday, they're asking for products by brand name. By three or three and a half, experts say, children start to believe that brands communicate their personal qualities, for example, that they're cool, or strong, or smart. Even before starting school, the likelihood of having a television in their bedroom is 25 percent, and their viewing time is just over two hours a day. Upon arrival at the schoolhouse steps, the typical first grader can evoke 200 brands. And he or she has already accumulated an unprecedented number of possessions, beginning with an average of seventy new toys a year.[66]

Complicit, wittingly or unwittingly, with a politics defined by market power, the American public offers little resistance to children's culture being expropriated and colonized by Madison Avenue advertisers. Eager to enthrall kids with invented fears and lacks, these advertisers also entice them with equally unimagined new desires, to prod them into spending money or to influence their parents to spend it in order to fill corporate coffers. Every child is vulnerable to the many advertisers who diversify markets through various niches, one of which is based on age. For example, the DVD industry sees toddlers as a lucrative market. Toy manufacturers now target children from birth to ten years of age. Children aged eight to twelve constitute a tween market and teens an additional one. Children visit stores and malls long before they enter elementary school, and children as young as eight years old make visits to malls without adults. Disney, Nickelodeon, and other megacompanies now provide Web sites such as "Pirates of the Caribbean" for children under ten years of age, luring them into a virtual world of potential consumers that reached 8.2 million in 2007, while it is predicted that this electronic mall will include 20 million children by 2011.[67] Moreover, as Brook Barnes points out in the *New York Times*, these electronic malls are hardly being used either as innocent entertainment or for educational purposes. On the contrary, she states, "Media conglomerates in particular think these sites—part online role-playing game and part social scene—can deliver quick growth, help keep movie franchises alive and instill brand loyalty in a generation of new customers."[68] But there is more at stake here than making money and promoting brand loyalty among young children: there is also the construction of particular modes of subjectivity, identification, and agency.

Some of these identities are on full display in advertising aimed at young girls. Market strategists are increasingly using sexually charged images to sell commodities, often representing the fantasies of an adult version of sexuality. For instance, Abercrombie & Fitch, a clothing franchise for young people, has earned a reputation for its risqué

catalogues filled with promotional ads of scantily clad kids and its over-
the-top sexual advice columns for teens and preteens; one catalogue
featured an ad for thongs for ten-year-olds with the words "eye candy"
and "wink wink" written on them.[69] Another clothing store sold
underwear geared toward teens with " 'Who needs Credit Cards...?'
written across the crotch."[70] Girls as young as six years old are being
sold lacy underwear, push-up bras, and "date night accessories" for
their various doll collections. In 2006, the Tesco department store
chain sold a pole-dancing kit designed for young girls to unleash the
sex kitten inside. Encouraging five- to ten-year-old children to model
themselves after sex workers suggests the degree to which matters of
ethics and propriety have been decoupled from the world of marketing
and advertising, even when the target audience is young children.
The representational politics at work in these marketing and advertising strategies connect children's bodies to a reductive notion of
sexuality, pleasure, and commodification, while depicting children's
sexuality and bodies as nothing more than objects for voyeuristic adult
consumption and crude financial profit.

Young boys and teens fare no better, as they are constantly being
initiated by Madison Avenue marketers into the world of hypermasculine, violent, and addiction-inducing programming and images. Video
games such as *Grand Theft Auto III*, Hollywood movies such as
*Alpha Dogs*, television programs devoted to extreme sports, contemporary wrestling, and other forms of media all provide entertainment
through a culture of bullying, homophobia, militarism, and misog-
yny. Moreover, they spin off a variety of products that embody such
values and practices, including everything from toys and magazines
to clothing. Even the military has learned from the playbook of the
architects of commercial culture by adopting the "culture of cool"
in its attempts to "create positive associations with the armed forces,
[immersing] the young in an alluring militarized world of fun,"[71] and
now supporting everything from rap concerts to video game cham-
pionships. Weaving itself into every aspect of popular culture, the
military-industrial complex now uses all elements of the media to spin
a positive view of militarism and violence as a hard-wired productive element of masculinity in order to promote military values as the
highest ideals in guiding young boys to manhood. Dressed up in the
aesthetic of the spectacle, the military-entertainment machine legitimates a hypercompetitive response in boys, nurturing and cultivating
the appeal of power, force, mastery, domination, and control through
the " 'masculine' virtues of toughness, strength, decisiveness, [and]
determination to stay the course while those who oppose such virtues

are 'feminized', [perceived as] sensitive, indecisive...weak, [that is], [t]hey are 'girlie-men.' "[72] This militarized worldview is now marketed to young children in the form of toys, clothing, cartoons, Hollywood films, video games, television programming, and other products and outlets. What is at stake here are not merely popular media representations of diverse forms of violence and masculinity but those broader institutions of political might, military power, and economic exchange that work in conjunction with those representations, offering mutual legitimation, bolstering influence, and securing widespread social acceptance, with the effect of contributing to the corruption of American childhood, and by extension democracy itself. Doug Kellner is worth repeating on this issue:

Yet, while media images of violence and specific books, films, TV shows, or artifacts of media culture may provide scripts for violent masculinity that young men act out, it is the broader culture of militarism, gun culture, extreme sports, ultra violent video and computer games, subcultures of bullying and violence, and the rewarding of ultramasculinity in the corporate and political worlds that is a major factor in constructing hegemonic violent masculinities.[73]

The militarized masculinity produced for boys and the degrading sexuality marketed to young girls complement and collude with each other in a society in which commodification reifies and fixates the range and complexity of possible identities young people may assume. Children are relentlessly and shamelessly exploited as fodder for prurient adults, military recruiters, and business types eager for financial gain. It might be a stretch to label such tactics as obscene or pornographic, but it is not an exaggeration to suggest that there is no place for such practices in a democracy that treats its children with dignity, compassion, and respect. In the end, what is clear is that modern advertising practices under the hypercapitalism of neoliberalism produce what can be called the swindle of agency and the denigration of civic responsibility.

Modern advertising, which cuts across all forms of traditional and new media, is relentless in its search for younger customers and its bombarding of young people incessantly with the pedagogy of commerce. While American children are being inundated with commercials, they are also spending more time with those technologies that deliver nonstop ads. Children have become a captive audience both to traditional forms of media, such as television and print, and to new media, such as mobile phones, MP3 players, the Internet, computers, and other forms of electronic culture that now seem to provide the latest products at the speed of light. Kids can download enormous amounts of media in seconds and carry around such information, images, and videos in a device the size of a thin cigarette lighter. Moreover, "[media] technologies themselves are morphing and merging, forming an ever-expanding presence throughout our daily environment."[74] Mobile phones alone have grown "to include video game platforms, e-mail devices, digital cameras, and Internet connections," making it easier for marketers and advertisers to reach young people.[75] Kids of all ages now find themselves in what the Berkeley Media Studies Group and the Center for Digital Democracy call "a new 'marketing ecosystem' that encompasses cell phones, mobile music devices, broadband video, instant messaging, video games, and virtual three-dimensional worlds," all of which provide the knowledge and information that young people use to navigate the consumer society.[76] The instructors and purveyors of commerce who control and service this massive virtual entertainment complex spend vast amounts of time trying to understand the needs, desires, tastes, preferences, social relations, and networks that define youth as a potential market. They not only employ sophisticated research models, ethnographic tools, and the expertise of academics to win over the hearts and minds of young people; they also work incessantly to develop strategies to deliver them to the market as both loyal consumers and commodities.

Time plays a particularly important role in the quest to turn young people into savvy shoppers and commodities. As young people of all ages become a captive audience for Madison Avenue advertisers, time in a consumer society is undermined as a public function as it becomes utterly privatized and corporatized. In this instance, there is a direct correlation between the rise of a consumer society under the biopolitics of neoliberalism and the transformation of public time into corporate time, especially with respect to the shaping of children's lives. Corporate time imagines time as units of labor, production, sales, and consumption, while focusing on short-term goals—largely defined by financial profit and the quick accumulation of capital.[77] Rather than following the script of public time, which decelerates the pace of everyday life in order to provide the noncommodified conditions and spaces for individuals and groups to think critically, debate, dialogue, and exercise judgment, corporate time is about marking time as instrumental. It is about reducing time to a measure of market considerations, profits, and financial exchanges, all of which is meant to promote the values suited to a consumer society. Corporate time rules children's lives and becomes a political deficit under conditions in which the most powerful business interests not only control the media but also insert children into a temporal world that captures "sixty or seventy hours a week, fifty-two weeks a year, of [their] time and attention."[78]

Children are increasingly exposed to a marketing and advertising pedagogical machinery eager and ready to transform them into full-fledged members of the consumer society. And the amount of time they spend in this commercial world is as breathtaking as it is disturbing. For instance, "It has been estimated that the typical child sees about 40,000 ads a year on TV alone,"[79] and that by the time they enter the fourth grade they have "memorized 300–400 brands."[80] In 2005, the Kaiser Family Foundation reported that young people are "exposed to the equivalent of 8½ hours a day of media content... [and that] the typical 8–18 year-old lives in a home with an average of 3.6 CD or tape players, 3.5 TVs, 3.3 radios, 2.0 VCRs/DVD players, 2.1 video game consoles, and 1.5 computers."[81] In the synoptic world of ads and marketing practices, the project of commercializing and commodifying children is ubiquitous and can be found wherever a previously noncommodified space existed. Hence, it comes as no surprise to find ads, logos, and other products of the marketing juggernaut pasted on school walls, public buildings, and public transportation systems, in textbooks and public washrooms, and even on baseball diamonds.

The influence of corporate culture is visible in the ongoing commercialization of public schooling, but its primary sites of public pedagogy are elsewhere, in a network of marketed cultures that permeates the everyday lives of both kids and adults. And what is disturbing about public time being lost in the sinkhole of commercialism is the concurrent disappearance of those crucial public spaces in which children are given the pedagogical skills to understand the social as a space and movement "in which the very question of the possibility of democracy becomes the frame within which a necessary radical learning (and questioning) is enabled."[82] As is well known, traditional modes of schooling and socialization are no longer the principle source of children's education, nor are they able to compete with these new and more effective tutors of "liquid modernity." This corporate matrix capitalizes on forms of public pedagogy that speak to kids outside of the school and have made commercial culture the most powerful force for educating children that the world has ever seen. Barber rightfully highlights this position in his comment:

Measured by the time allotted to them, commercialism's pedagogical competitors—education, parenting, socialization by church or civic groups—come out on the short side. Teachers struggle for the attention of their students for at most twenty or thirty hours a week, perhaps thirty weeks a year, in settings they do not fully control and in institutions that are often ridiculed in the popular media.... the true tutors of late consumer capitalist society as measured by time are those who control the media monopolies, the aggressive content purveyors, shameless lords of the omnipresent pixels, who capture sixty or seventy hours a week, fifty-two weeks a year, of children's time and attention.[83]

Of course, young people are not just watching and passively inhabiting this floating and endlessly growing world of products, ideas, and values; they are also learning how to mediate and define themselves within this parade of sales pitches, ads, and commodified spectacles. But the spaces necessary for them to offer both resistance and a challenge to the pedagogical machinery of neoliberalism are disappearing.[84] Market values are being extended beyond the workplace into the wider society, and neoliberalism is developing a promotional culture that "mobilizes thinking, imagination and sensibility as businesses attempt to capture customer loyalty.... develop new markets, [and win over] the intimacy of the consumer in order to embed commercial transactions in personal and daily life."[85] The high priests of marketing and advertising defend their pedagogical influence on children's minds, consciousness, and culture by claiming that they are empowering kids by giving them more choices. But Juliet Schor finds that underlying the relentless commercialization of children is a marketing blitz aimed at defining children in terms of economic potential, one that relies on the much more revealing and corrupting metaphors of war. She writes:

One clue to the marketing mentality is industry language. It's a war out there. Those at whom ads are directed are "targets." When money is committed to an ad campaign it is referred to as "going against the target." Printed materials are called "collateral." Impromptu interviews with consumers are "intercepts." The industry is heavily into the metaphor of biological warfare, as in the terms "viral marketing" and "sending out a virus."... There's not much doubt about who's winning this war either. When Nickelodeon tells its advertisers that it "owns kids aged 2-12," the boast is closer to the mark than most of us realize.[86]

## THE PEDAGOGY OF COMMODIFICATION

The potential for lucrative profits to be made off the spending habits and economic influence of kids has certainly not been lost on a small number of megacorporations, which under the deregulated, privatized, no-holds-barred world of neoliberalism have set out to embed the dynamics of commerce, exchange value, and commercial transactions into every aspect of personal and daily life. If these corporations had their way, then kids' culture would become not merely a new market for the accumulation of capital but a petri dish for producing new neoliberal subjects. As a group, young people are vulnerable to Madison Avenue advertisers who make every effort "to expand inwardly into the psyche and emotional life of the individual in order to utilize human potential" in the service of a market society.[87] Since children's identities have to be actively directed toward the role of consumers, knowledge, information, symbolic meaning, and learning become central in shaping and influencing every waking moment of children's daily lives. In this instance, corporations, marketers, and advertisers are not just exploiting kids for profit; they are actually both constructing them as commodities and promoting the concept of childhood as a saleable commodity. As Schor points out,

Marketers are scrutinizing virtually every activity kids now engage in—from playing, eating, and grooming, to bathing—and virtually every aspect of their lives—from what's inside their closets to how kids interact in the classroom and what really goes on at a tween girl's slumber party. They are probing what kids talk about, and even how they use drugs.... They are influential in actually producing children—that is, in raising, educating, forming, and shaping them. And they do this in a commodified form; that is, they produce children in order to sell them back to their clients. They create in-depth research that they then sell. They provide children with cultural products such as television programming, movies, and web content. They sponsor museum exhibits, school curricula, and leisure activities for children, all of which help to create children as social beings. Advertisers have even gotten into the business of structuring the form and content of social interaction and conversation among children, a phenomenon they benignly term "peer-to-peer marketing." In the last fifteen years, advertisers and marketers have been extraordinarily successful in these endeavors.[88]

What is particularly disturbing in this scenario is the growing number of marketers and advertisers who work with child psychologists and other experts who study children in order to better understand children's culture so as to develop marketing methods that are more camouflaged, seductive, and successful.[89] This is evident in a variety of marketing techniques in which kids are co-opted by corporations. Using various inducements from money to products to free tickets, corporations now enlist kids both to test products and supply information about what is "hot" among trendsetters. More specifically, they "use gangs of 'cool' children to push products in their peer groups and communities, exploiting the power of peer pressure and children's fears of not fitting in for profit-making."[90] Given the power young people have to influence the spending habits of their parents, corporations have developed a number of strategies that encourage children to pester them to buy sought-after merchandise. By constantly bombarding kids with messages about what is cool, trendy, and available, corporations set the stage for encouraging kids to nag their parents on trips to the supermarket, the mall, and other venues that offer children's products. For example, during the holidays, especially Christmas, many retailers such as Wal-Mart use their Web sites to provide wish lists for children. "[C]hildren are asked to pick items from a conveyor belt and build up a wish list of items that, conveniently, are all available in Wal-Mart stores. The web site then encourages children to enter their parents' e-mail addresses so it can send them the list and 'help pester your parents for you.' "[91] Within this pedagogical template, parents are useful only as a potential source of goods for kids and profits for corporations. In fact, central to the pedagogy of commodification is the notion that "adults are never cool—they are boring, often absurd, sometimes stupid—and when they try to be cool they are pathetic."[92] There is more at work in this strategy than anti-adultism; there is a shrewd attempt to open a space free of adults that marketers and advertisers can tap into without having to deal with the mediating influence of the adult gaze. There is also an attempt to replace parental authority with the authority of the market, especially when it comes to the matters of commerce and commodification. The Nickelodeon motto "Kids Rule" has nothing to do with either empowering kids or taking their voices seriously. On the contrary, it represents both a departure from the panoptic, disciplinary obsessions of traditional institutions of socialization and a new attempt to refashion "pastoral power" and guidance for children under the omnipotent gaze of advertising, rather than religious institutions, schools, and family.[93]

Anti-adultism entails, borrowing from Michel Foucault, a new type of sovereignty and mode of governmentality that provide a novel understanding of power, signaling a new kind of "encounter between the technologies of domination of others and those of the self."[94] Governmentality is less about the legitimation of state institutions than it is about the regulation of consent, persuasion, and the harnessing of human energy through the use of technologies that privilege pedagogy and the production of specific forms of subjectivity. Governmentality involves, as Arjun Appadurai puts it, "the large scale coordination of persons, resources, and loyalties."[95] As a way of both managing youth and producing youth as reliably consuming subjects, marketers now bypass parents by using a pedagogy of commodification that directly addresses kids in their schools, homes, leisure activities, and places of work while at the same time analyzing virtually every activity in which kids engage. At the same time, it is important to stress that adults are increasingly separated from children and the spaces they occupy, leaving them largely unprotected from the ravages of the market. In addition, adults themselves are increasingly interpolated and folded into a system of high consumption, thus contributing to the educating of children as consumers.[96]

As young people become the objects of increasing anthropological studies, corporations mine every aspect of their lives and diverse modes of cultural production—from music to clothing to hairstyles—for ideas that can be translated into a source of profit. Entire companies are now organized to do research on young people, gather information, and sell it to client corporations, making it all the more difficult for children and youth to resist the ever-growing and changing ads, products, and ideas aimed at turning kids into commodities. With approximately $200 billion in spending power, young people now represent an eagerly sought-after market. They also become part of what Alex Molnar and Faith Boninger call the "total environment" of marketing, which provides the most pervasive and persuasive sources of knowledge for young people.

In 2007, we see a marketing environment that recognizes few boundaries. Advertisers ply their trade wherever they can and even engage consumers as collaborators in their marketing strategies. This "total environment" of marketing is enabled in part by new technologies that allow advertisements to appear in places they could not have been before, such as video games, social networking web sites, and cell-phones. It is also the result of greater cultural acceptance of marketing as an everyday fact of life, a friendly political environment, and a willingness on the part of marketers and advertisers to breach boundaries that previously limited their activities. Whereas, for example, there used to be a clear boundary between "editorial content" (e.g., television programming, magazine articles, or school curricula) and advertisements, we now see the judges on American Idol sipping from Coca-Cola cups, the debonair cavemen from Geico commercials starring in their own television program, and Disney Publishing providing comics to schools for a reading program.[97]

When young people are not being scrutinized, they are enlisted by corporations to pose as consumers in order to promote products in bars and other social settings. Recognizing the power of teens to influence their peers, marketers aggressively recruit young people to boost products, establish markets from among their friends, and initiate them into a set of stealth marketing methods in which they learn that all social relations, even friendships, are now largely defined as potentially lucrative sources of profit. In this network of commercial values and commerce, social relations become utterly instrumentalized. Carly Stasko and Trevor Norris point to the reach and power of this type of marketing with the following example: "Procter & Gamble recently established a subsidiary called 'Tremor,' which has 'assembled a stealth sales force of teenagers—280,000 strong—to push products on friends and family.' "[98] In addition, marketing and advertising firms such as Mr. Youth, YMS Consulting, EPM Communications, the Gaped Group, and the Girls Intelligence Agency (GIA) do everything from providing advice on marketing to young people and keeping corporate clients aware of new developments in the youth market to recruiting young people to serve as a resource for corporations who want to create a "buzz" for their products.[99] YMS's mission statement is a particularly clear expression of a business venture that employs the rhetoric of childhood commodification with no apparent ethical misgivings about selling children as markets to corporations, asserting (without irony) that they enrich the lives of young people. It states:

Our mission is to be the most up to date and insightful "Hub" of Information, Expertise and Advice for all those who seek to responsibly market to today's youth and their families. Simply put, our goal is our client's success—especially when it results in the enrichment of the lives of the young people and their families who become the consumers of our clients' products and programs.[100]

Kids are now hired to use their friends to gain information about what is cool and trendy, information that is later used to sell products to a larger teen demographic. Within this type of pedagogy, kids are not merely commodified. They also learn quickly that duplicity and deception are perfectly acceptable in a world in which manipulation translates into free products such as films, toys, and concert tickets. In one particularly insidious practice that is emblematic of this type of marketing in which friends now become valuable as a potential source of money, products, or status, GIA sponsors what they call "Slumber Party in a Box." In a description of the event posted online by GIA headquarters, readers are told that "GIA Slumber Party in a

Box is a customized party kit [that] includes hot, new products that are pro-girl." Selected agents invite their friends to an overnight party, hand out free products to them, and then provide "feedback through quizzes" to GIA headquarters. Mimicking a page from the playbook of the CIA, to which it owes a debt for more than its tacky moniker, GIA makes secrecy crucial to the success of the event, since the agents don't have to reveal to their friends that the "party" is underwritten by a marketing firm. Interestingly enough, GIA instructs the girls "to be slick and find out some sly scoop on your friend," such as what products they are interested in, what is considered cool, and so on.[101] There is more than manipulation, deception, and disingenuousness at work here; there is also the erosion of civic ideals and the furthering of a currently fashionable social Darwinism. The twist here is that it is now children themselves determining who and what is cool and acceptable and who is uncool, deserving of ridicule or social exclusion. Bauman describes it well: "They all tell us the same story: that no one is of use to other human beings [unless] she or he can be exploited to their advantage, that the waste bin, the ultimate destination of the excluded, is the natural prospect for those who no longer fit or no longer wish to be exploited in such a way, that survival is the name of the game of human togetherness and that the ultimate stake of survival is outliving the others."[102]

The pedagogy of commodification takes many forms, but what these different marketing methods all share is both an unqualified belief in childhood agency as a mix of self-promotion and consumption and a willingness to render disposable those young people who cannot fulfill the mandate of acceptable consumer habits. The threat of social exclusion is the method of choice for mobilizing fear in potential youthful consumers. Marketing concepts such as "cool" operate off the assumption that social relations work primarily as a site of intense competition, pitting youngsters who are trendy against those who cannot keep up within an ever-changing economy of objects and fashions, undergoing as a first principle instant obsolescence. Kids are under intense pressure to keep up with trends, learning quickly that under the regime of market sovereignty their value, if not their dignity and identity, rests on what they accumulate rather than who they are. With alarming candor, Nancy Shalek, president of a reputable advertising company, revealed one element of the neoliberal ideology that drives her pursuit of the youth market. She claims without apology that "advertising at its best is making people feel that without their product, you're a loser. Kids are very sensitive to that. If you tell them to buy something, they are resistant. But if you tell them that they'll be a dork if they don't you've got their attention. You open up emotional vulnerabilities and it's very easy to do with kids because they're the most emotionally vulnerable."[103] Kids learn early that social exclusion is tied to one's status as a shopper, that identities are determined by owning the most fashionable commodities, and that their childhood and future success depends on their role as consumers. This is a task that becomes more and more difficult as an increasing number of kids become part of a "Recession Generation"— teetering on or slipping into poverty.[104] Not being in vogue, being out of synch with the images of perfection provided by advertisers, or being uncool when it comes to defining oneself as a commodity suggest a social order that involves more than a consumer lifestyle. Kids are living in an all-embracing consumer society that produces consumer identities in which many kids are not just ensnarled but also locked out of any viable notion of happiness and agency unrelated to their ability to buy. More pointedly, the underside of being cool is being relegated to the landscape of shame and rejection, while legitimating a world in which inequality provides the framework for rewarding elitist and exploitative notions of identity and social status. Children learn quickly that as a flawed consumer one stands little chance of success in a consumer society. But, even more importantly, as Bauman points out, "[c]hildren had better start bracing themselves early for the role of eager and knowledgeable shoppers/consumers— preferably from birth. No money spent on their training will be money wasted."[105] What becomes clear in this pedagogy of commodification, anti-adultism, and social expulsion is that it does more than expose children to unnecessary feelings of fear, vulnerability, and humiliation as it attacks their sense of self and their capacity to bond with others. It also demonstrates that adults are unwilling in this corporate world order to assume the necessary responsibility in protecting children from such outrages. In spite of what some critics have argued, childhood at the beginning of the twenty-first century is not ending as a historical and social category; instead, it has simply been transformed into a market strategy and a fashionable aesthetic used to expand the consumer-based needs of privileged adults who live within this devalued Gilded Age that has little concern for ethical considerations, noncommercial spaces, or public responsibilities.[106] What is changing, if not disappearing, are productive social bonds between adults and children, and even among youngsters themselves.

The price of emotional vulnerability comes high for young people in a market society. But the pedagogical values, conditions, and processes that make it possible are sharply revealed through the corporate practice of branding. While Naomi Klein introduced a critique of the concept of branding to a public audience in her book *No Logo*, the concept had a very long and storied history and reception in the world of commerce. For instance, *Fast Company*, a business-friendly magazine, in 1997 devoted an entire issue to the positive qualities of branding, making clear that the commodification of the self was no longer something to be ashamed of. Seizing upon a neoliberal rhetoric that had become normalized, Tom Peters, one of the founders of what has been called the "management guru industry," grandly announced that "[r]egardless of age, regardless of position . . . all of us need to understand the importance of branding. We are CEOs of our own companies: Me Inc. . . . Our most important job is to be head marketer for the brand called You."[107] Peters was unequivocal in his belief that the most important measure of success in life was determined by how successfully individuals turned themselves into brands. As he put it, "Starting today you are a brand. You're every bit as much a brand as Nike, Coke, Pepsi, or the Body Shop. To start thinking like your own favorite brand manager ask yourself . . . What is it that my product or service does that makes it different?"[108] Peters believed that to get ahead people had to market themselves by getting a "marketing brochure for brand You." And, of course, hubris, greed, or narcissism should not get in the way since "Being CEO of Me Inc. requires you to act selfishly—to grow yourself, to promote yourself, to get the market to reward yourself."[109] Peters spoke unabashedly to what had become a cardinal principle of everyday life: one's personal life and social relations now had to be fashioned out of the tools and values provided by a market society. The second principle of branding spoke more directly to the decoupling of notions of self and agency from "specific social roles and obligations" and to their redefinition almost exclusively as "privatized self interests and desires," insisting that one's understanding of the world should be mediated primarily through the marketing and consumption of goods, services, knowledge, and images.[110]

Branding has played an enormous role in convincing generations of young people that instead of simply buying goods, they were buying lifestyles, worldviews, ideas, and images. Goods could now be marketed as dreamscapes of desire, liberatory experiences, and modes of power. Rather than simply buy commodities, adults and children could now become the commodity by owning it, attaching status-oriented logos to clothing and actually promoting their identities as a commodity. Everywhere children look, brands stand in for substantive identities. In the world of sports, many of the star athletes greatly admired by young people become indistinguishable from the products they promote. Michael Jordan, one of the most famous basketball players in the world, pitched everything from Gatorade to Nike products. Tiger Woods, the world's most recognizable golf star, is now a marketing spokesperson for Apple Computers, Buick, and a host of other companies. Rather than using their celebrity status for educating young people about character, hard work, the value of sportsmanship, and the sheer joy of athleticism, these athletes deceive young people into believing that becoming the embodiment of a brand is the apogee of what it means to be a successful adult in the world of sports and that childhood is largely the training ground for the eventual selling of the self. Young women are similarly surrounded by famous females whose talents are outstripped by their brand status, at least according to the media. It is precisely under such conditions that the well-regarded *New York Times Sunday Magazine* can, without irony or sustained critique, run a cover story that reads more like an advertisement, lauding Tyra Banks, the supermodel, with the title "Bankable: How Tyra Banks Turned Herself Fiercely into a Brand."[111] Clearly, branding has a cachet that even powerful newspapers are willing to exploit regardless of the message it sends to children.

Of course, the pedagogy of branding is not all of one piece. Some of it is quite subtle, and some of it is simply shameless. One particularly egregious example appeared in a story that the *New York Times* ran in 2008 on Floyd Mayweather, considered by many experts to be pound for pound the best boxer in the world. It seems that Mayweather, though one of the most recognized boxers in the world, wants more recognition. Hence, the reader is informed that he is changing his first name to "Money" and is hell-bent on increasing his brand status. The *Times'* description of Mayweather's projected evolution to an "A-lister" is worth repeating in full.

"Boxing is Floyd's platform, but it's not a mainstream sport anymore," says Leonard Ellerbe, his manager. "To get into the mainstream, you have to do mainstream things." Mayweather has "elevated the brand and expanded the fan base" and become "an A-lister," as Ellerbe puts it, not only by winning all of his professional fights and earning a fortune ($50 million in 2007 alone), but also by dancing with the stars, palling around with 50 Cent and Mark Cuban, starring in a reality show on HBO, rapping, venturing into music production, promoting concert tours by Beyoncé and Chris Brown, waving the green flag at the Indianapolis 500 and appearing on TV talk shows. Ellerbe promises that more such dabbling is on the way, including movie deals and a "stimulation beverage." It is all part of what Mayweather and Ellerbe both refer to as Mayweather's "ultimate goal" of turning himself into "a Fortune 500 company" and becoming "the biggest entertainer in entertainment"—kind of like what the Joker has in mind in "Batman" when he says he wants his face on the dollar bill.[112]

Branding is the ultimate expression of the pedagogy of commodification, and its disregard for children's daily experiences and potential future is on full display. Not only does it locate them within an emotional landscape and political economy that translate the act of citizenship into the art of consuming; it also undermines any sense of viable agency they might assume in a putatively democratic society by turning their needs, desires, and hopes over to a legion of corporate vampires. As part of the broader logic of a biopolitics of neoliberalism, branding instills in young people a deep sense of self-delusion and doubt, while decoupling this usurpation of dignity from the broader realms of the social and political. For many young people, branding produces a sense of humiliation and indignity; children who rank among the poor and who are failed consumers cannot participate in its socially constructed rituals and rewards. For too many children, the only outlet for understanding or contesting such practices is a mode of resentment that is both depoliticized and individualized.

As part of the pedagogy of commodification, branding has become an anesthetic, the political equivalent of ether, wrapped up in the discourse of empowerment and choice. Branding shifts the loyalties of children away from their parents toward corporations. Moreover, it gains growing legitimacy and faces little opposition from either young people themselves or adults, as democratic public spheres capable of producing a language of critique, justice, and hope disappear. Schools are turned into shopping malls; public spaces are transformed into commercial spaces; and the discourse of freedom is reduced to the mantra of consumer choice, which translates into the freedom of choice to consume endlessly and dispose of goods, services, and relationships just as quickly.[113] Choice within this commodified notion of freedom comes without the need to worry about consequences, since living one's life as a commodity means allowing the market and its avatars to make the choices that deeply affect one's sense of agency. Freedom as a form of disempowerment becomes even more obvious as young and old alike are informed under the privatizing rhetoric of the market that "the responsibility for choices, the actions that follow the choices and the consequences of such actions rests fully on the shoulders of individual actors."[114] Branding empties young people

of moral ideals while promoting the unchecked, rampant commodification, commercialization, and privatization of children's culture
and lives. Within this discourse, young people are called upon to
deal with their lack of self-confidence, powerlessness, and the endless
indignities heaped upon them in a consumer society by employing
utterly privatized and individualized solutions to what are "socially
produced discomforts."[115] The result is that they can locate the source
of their troubles only in themselves, thus translating their perceived
insufficiency into both envy for the rich and disdain for themselves.[116]

Susan Gal argues that "[t]he strongest form of power may well be
the ability to define social reality, to impose visions of the world."[117]
This is particularly true for the kind of political and pedagogical power
wielded by huge corporations that target young people, who, in spite
of their ability to mediate those forces that shape their lives with
some degree of critical insight, are emotionally vulnerable and largely
powerless in the face of such corporate strength. Steven Heller, in
his book *Iron Fists*, goes further and argues that corporate branding
strategies used to guarantee consumption and commodify subjects are
an insidious form of propaganda, not unlike that adopted by some
of the most destructive twentieth-century totalitarian regimes as a
way of maintaining a grip on their populations.[118] Sheldon Wolin
sees an eschatological connection among modern advertising, religious fundamentalism, and authoritarian regimes. He writes, "Equally
important, the culture produced by modern advertising, which seems
at first glance to be resolutely secular and materialistic, the antitheses
of religious and especially of evangelical teachings, actually reinforces
that dynamic. Almost every product promises to change your life: it
will make you more beautiful, cleaner, more sexually alluring, and
more successful. Born again, as it were... Each colludes with the
other. The evangelist looks forward to the 'last days,' while the corporate executive systematically exhausts the world's scarce resources."[119]
Surely the totalitarian logic and social death implicit in the branding of
an entire generation of young people are on full display in "neuromar-
keting," which uses "magnetic resonance imaging (MRI) to map brain
patterns and reveal how consumers respond to a particular advertisement or product."[120] There is more at work here than the insidious
practice of using a neurological technology that is crucial for medical
purposes to map people's desires in the interest of corporate profits;
there is also the resurgence of an instrumental rationality whose end
point will not be the mapping and manipulation of people's desires,
but the far more ominous prospect of the elimination of redundant
and flawed consumers.

Under the biopolitics of neoliberalism, the commodification of young people cannot be separated from the social totality and a new mode of sovereignty in which free-floating global economic forces have "the power and capacity to dictate who may live and who may die."[121] The logic of disposability as it works through the politics of commodification has a direct bearing on the lives of young people and, therefore, on the future health of democracy. For example, the marketing of junk food and other products that promote health hazards coupled with the collapse of the social state and diminishing health services, the elimination of basic social provisions for children, and the increased commercial pressures kids experience in a consumer society all undermine the well-being and lives of young people. In addition, the neoliberal logic of disposability further legitimates eliminating those public spheres capable of providing children with an education that gives them access to valuable intellectual resources and nurtures a critical sense of agency—all of these factors render the fate of children and the prospect of future democracy deeply unstable and foreboding. For instance, children are inundated by media advertising campaigns that promote unhealthy foods such as candy, soda, and snacks while also being held hostage to fast-food chains that now, because of the push to privatize and corporatize public schools, have a solid foothold in the American educational establishment. Consequently, children are suffering from a range of serious health problems including high blood pressure, increased cholesterol levels, heart problems, Type 2 diabetes, hypertension, obesity, respiratory ailments, and a near epidemic of stress—all of which are exacerbated by the sedentary lifestyle pitched to kids by a consumer society.[122] The pressures young people are facing in a society that simultaneously attacks their sense of security and self-esteem are evident in the record levels of emotional problems young people are experiencing, ranging from depression and esteem issues to high levels of anxiety and social dysfunction. All of these are compounded by the subjection of millions of children to abusive forms of medicalization and hospitalization.

Even some in the dominant media have begun to acknowledge the unprecedented burden placed on kids today as a result of neoliberal ideology and a market fundamentalism largely driven by the need for new consumers and the lure of immense profits. For example, Judith Warner, a writer for the *New York Times*, argues that young people live in a society that suffers both from "a social pathology in which getting and spending...have become our nation's most cherished activities [leading to] waste and harm to the environment" and from a host of problems directly affecting kids, such as "teen suicide, pregnancy, substance use and violence, [increased] rates of cigarette smoking, depression, and alcohol problems."[123] Even public schools, which have a civic responsibility to nurture the critical intellect, social responsibilities, and civic capacities of young people, have been colonized by various corporate industries. Not only do they influence student diets, but they also inundate kids with commercialized curricula and relentless ads that undermine their physical well-being and mental health. For example, Channel One Network gets inside of public schools by offering financially strapped districts borrowed television equipment in exchange for mandatory daily viewing by students of twelve minutes of soft news and advertisements. It has been estimated that the "Channel One News reaches more than 7 million secondary-school children in 11,000 schools," an audience that is 50 times larger than the MTV teen audience.[124] In a neoliberal regime, basic social institutions that children rely upon are starved financially in the exact proportion that commercial enterprises are given unprecedented access to kids, appropriating and distorting the role of trusted mentor and teacher.

What is being taught by a current popular educational reform movement that embraces a market mentality by paying kids with cash rewards if they perform well enough to raise standardized test scores? The most persuasive lessons really being taught in this market-oriented pedagogical practice are that financial reward provides the only motive that counts in life and that the worth of everything from knowledge to social relations is ultimately measured by its exchange value. There is a buried order of politics at work in this value system, one that strongly aligns itself with a market-based rationality that defines freedom exclusively through the acquisition of money and wealth, and human values exclusively as a token of economic exchange. This "motivation for sale" pedagogy has no language for relating the self to public life, to social responsibility, or to the demands of citizenship. Where are children to find the knowledge, values, and spaces that enable them to recognize antidemocratic forms of power, develop critical modes of literacy, fight substantive injustices in a society founded on deep inequalities, and embrace the broader complex of hope and vision that keeps the promise of democracy alive? Certainly not in many schools, in which the concept of investment is decoupled from the common good and becomes a marketing tool and pedagogical ploy for young people to define themselves entirely through the logic of the market-driven ideology of neoliberalism.

There is yet another type of disposability at work in consumer culture—the erosion of the childhood imagination, the slow death of the ability of young people to imagine a future outside of the lure of commodities, and the wearing away of any viable notion of human solidarity, social responsibility, and critically engaged citizenship. As the public domain is transformed into the ruins of a homogenizing "corporate monoculture," debate, critical dialogue, dissent, judgment, and thoughtfulness, all of which are central to any substantive democracy, become suspicious, if not scorned, activities.[125] Under the biopolitics of neoliberalism, capitalism has taken a deadly turn, inculcating young people with material values that not only undermine civic life but present them with a notion of agency that is utterly privatized and at odds with democratic notions of the public good and freedom. And the impact that this pedagogy of commodification is having on young people does injury to their sense of agency, while boding ill for the future of democracy itself. As Schor points out, young people are utterly transformed through the corporate assault on their identities. She writes:

[They are] the most brand-oriented, consumer-involved, and materialist generation in history. And they top the list globally. A survey of youth from seventy cities in more than fifteen countries finds that 75 percent of U.S. tweens want to be rich, a higher percentage than anywhere else in the world except India, where the results were identical. Sixty-one percent want to be famous. More children here than anywhere else believe that their clothes and brands describe who they are and define their social status. American kids display more brand affinity than their counterparts anywhere else in the world. Indeed, experts describe them as increasingly "bonded to brands."[126]

If more and more young people in the United States and other developed countries are "bonded to brands," they are also held as a captive audience to the commodified landscapes they inhabit. One of the most insidious consequences is ethical and political indifference to the fact that the world of commodities and the conditions that produce their own indulgence in commodification arise, in part, from the terrible suffering and exploitation of children and youth globally. These are the children who work in sweatshops under horrendous conditions and who are among the main producers of the commodities and disposable goods that overflow the junkyards of the consumer society.

## CONCLUSION

Everything that can be done to bring the age of heroic consumption to its close should be done. This means the promotion of a different understanding of wealth. The myriad aspects of a truly rich and fulfilled life should be rescued from the tyranny of money.... This is why the A-listers, the celebs, the fat cats, the big spenders, the conspicuous consumers do not represent a "lifestyle" to be emulated at all costs, but serve as warning of the spectre of depletion and exhaustion awaiting us within a short space of time.[127] Jeremy Seabrook, "Consuming Passions: Everything That Can Be Done to Bring the Age of Heroic Consumption to Its Close Should Be Done"

For the last few decades, critics such as Thomas Frank, Kevin Phillips, David Harvey, and many others have warned us, and rightly so, that right-wing conservatives and neoliberals have been dismantling government by selling it off to the highest or "friendliest" bidder. But what they have not recognized adequately is that what has also been sold off are both our children and our collective future, and that the consequences of this catastrophe can only be understood within the larger framework of a biopolitics and market philosophy that view children as commodities and democracy as the enemy. In a democracy, education is utterly adverse to treating young people as individual units of economic potential and as walking commodities. But of course, we are motivated to "forget" that democracy should not be confused with a hypercapitalism. That is, what must be clarified as part of any democratic project is that consumption alone, as opposed to unbridled consumption, is not the enemy, just as the crisis youth face today should not lead to simply condemning all acts of consuming indiscriminately. We should not accept a totalizing critique of consumerism that refuses to rescue the utopian moments locked in the commodity. Inevitably, humans must consume to survive. The real enemy is a consumer society fueled by the endless cycle of acquisition, waste, and disposability, which is at the heart of global neoliberalism. But is there no remaining space in which to imagine a mode of consumption that rejects the logic of commodification and embraces the principles of sustainability while expanding the reach and possibilities of a substantive democracy? Juliet Schor makes this a central issue by asking "what kind of consumers do we want to be?"[128]

Although I agree with searching for an answer to this question, I also think that it only becomes meaningful when connected to the broader issue and question: what kind of society and world do we want to live in? Such a question suggests that any notion of commodification has to be understood and challenged within a broader political, educational, and economic project as a form of biopolitics that raises fundamental questions about the democratic nature of sovereignty, governance, power, and the everyday lives of our children. As politics embraces all aspects of children's lives, it is crucial to make clear that the rising tide of free markets has less to do with ensuring democracy and freedom than with spreading a reign of terror around the globe, affecting the most vulnerable populations in the cruelest of ways. The politics of commodification and its underlying logic of waste and disposability do irreparable harm to children, but the resulting material, psychological, and spiritual injury they incur must be understood not merely as a political and economic issue but also as a pedagogical concern.

The neoliberal attack on the social state and its disinvestment in those political, social, and economic spheres vital to the development of healthy and critically informed young people work in tandem with a new and vicious public pedagogy produced in countless sites that constructs youth according to the dictates, values, and needs of a market fundamentalism. It is also part of a culture of so-called security, force, and violence that produces wars abroad and a less intense but equally debilitating war at home. War in this case operates not only within the parameters of state power but also within the principles of market sovereignty. Simply criticizing the market, the privatization of public goods, and the commercialization of children, while helpful, is not enough. Stirring denunciations of what a neoliberal society does to kids do not go far enough. What is equally necessary is developing public spaces and social movements that help young people to transform themselves into engaged social actors. Jacques Rancière touches on this issue with his insistence that "[t]he critique of the market today has become a morose reassessment that, contrary to its stated aims, serves to forestall the emancipation of minds and practices. And it ends up sounding not dissimilar to reactionary discourse. These critics of the market call for subversion only to declare it impossible and to abandon all hope for emancipation."[129] Rancière cannot imagine a mode of criticism or a politics that shuts down resistance, play, and hope—nor should we as teachers, parents, and young people.

At stake here is the need for a new global politics of resistance and hope that mounts a collective challenge to a ruthless market fundamentalism that is spearheading the accumulation of capital, the commodification of young people, and the usurpation of democratic modes of governance. At the center of this struggle is a market sovereignty that has replaced the state as the principal regulatory force in developing economies of inequality and power that gain legitimacy and strength through modes of governmentality, persuasion, and consent that rely on the force of new technologies, pedagogical practices, and a calculating rationality, all of which have to be challenged and transformed. Any politics capable of disabling the sovereignty of the market must clarify in political and pedagogical terms a vision, project, discourse, and set of strategic practices necessary to confront a neoliberal order that views democracy as the enemy and youth as expendable. Neoliberalism has played a major role in creating not only widening class disparities but also a weakened social state and a failing democracy, made all the more ominous by the dumbing down of public discourse and the emptying out of public spheres.

Democracy is about neither the sovereignty of the market nor a form of state governance based largely on fear, manipulation, and deceit. Challenging neoliberal sovereignty and the national (in)security state means recognizing the need for a politics in which matters of education, power, and governance are mutually determined. Such a challenge, in part, rests on a politics that seeks to understand how governmentality and the pedagogy of commodification are produced and circulated through new modes of market sovereignty. Democratic politics and the struggles informed by such a politics cannot come about without putting into place the spaces, spheres, and modes of education that enable people to realize that, in a real democracy, power has to be responsive to the needs, hopes, and desires of citizens and other inhabitants. Democracy is not simply about people wanting to improve their lives; it is more importantly about their willingness to struggle to protect their right to self-government in the interest of the common good. Sheldon Wolin has rightly argued that "[i]f democracy is about participating in self-government, its first requirement is a supportive culture of complex beliefs, values, and practices to nurture equality, cooperation, and freedom."[130] The carceral state and the sovereign market reduce the materiality of democracy to either an overcrowded prison or a shopping mall, both of which are more fitting for a society succumbing to the impulses of totalitarianism.[131] They hollow out the institutional and educational conditions that conceive of social, political, and personal rights as fundamental to any viable notion of agency. As education turns to training in the public schools, higher education willingly models itself as a business venture, and the wider corporate culture becomes the most powerful pedagogical force in the country, while "democracy becomes dangerously empty."[132] As I argue in the concluding chapter of this book, we need a new politics that makes education central to the struggle for democracy, one that is not limited to schools and that is willing to enlist and actively mobilize artists, intellectuals, academics, and others in the struggle for a public able and willing to confront a reactionary state and a consumer society through multiple levels of resistance. Bauman illuminates this issue with razorlike precision:

Adverse odds may be overwhelming, and yet democratic (or, as Cornelius Castoriadis would say, an autonomous) society knows of no substitute for education and self-education as a means to influence the turn of events that can be squared with its own nature, while that nature cannot be preserved for long without "critical pedagogy"—education sharpening its critical edge, "making society feel guilty" and "stirring things up" through stirring human consciences. The fates of freedom, of democracy that makes it possible while being made possible by it, and of education that breeds dissatisfaction with the level of both freedom and democracy achieved thus far, are inextricably connected and not to be detached from one another. One may view that intimate connection as another specimen of a vicious circle—but it is within that circle that human hopes and the chances of humanity are inscribed, and can be nowhere else.[133]

Making education central to any viable notion of politics as well as making the political more pedagogical suggests that intellectuals, artists, community workers, parents, and others need to connect with young people in those public and virtual sites and spheres that not only enable new modes of dialogue to take place but also work to move beyond such exchanges to the much more difficult task of building organized and sustainable social movements. It is true that anyone who takes politics seriously needs to consider the profound transformations that have taken place in the public sphere—especially those enabled by new technologies—and how such changes can be used to develop new modes of public pedagogy in which young people are provided with the skills, knowledge, interests, and desire to govern themselves, but it is simply wrong to suggest that real change happens only online.[134] Building a more just, ecologically sustainable, and democratic future, or as Jacques Derrida put it, the promise of "a democracy to come,"[135] demands a politics in which the new technologies are important but only insofar as they are used in the context of bringing people together, reclaiming those public spheres where people can meet, talk, and plan collective actions. We must learn to resist all technologies that reinforce the sense of excessive individualism and privatization at the heart of the neoliberal worldview. In other words, we need a politics that reinvents the concept of the social while providing a language of critique and hope forged not in isolation but in a collective struggle that takes social responsibility, commitment, and justice seriously.

We live at a time when social bonds are crumbling and institutions that provide collective help are disappearing. Reclaiming these social bonds and the protections of the social state means, in part, developing a new mode of politics and pedagogy in which young people are as central to this struggle as the future of the democratic society they once symbolized. At the heart of this struggle for both young people and adults is the pressing problem of organizing and energizing a vibrant cultural politics to counter the conditions of political apathy, distrust, and social disengagement so pervasive under the biopolitics of neoliberalism. Material relations of power cannot be abstracted from symbolic power and the necessity to connect and reclaim the crucial relationship between political agency and democracy. For this we need a new vocabulary that, in part, demands taking back formal education and diverse modes of public pedagogy for democratic purposes, while also refashioning social movements and modes of collective resistance that are democratic in nature and global in reach. Culture in this instance is not merely a resource, but an instrument of political power. What must be emphasized in this vision of a democracy to come is that there is no room for a politics animated by a rationality of maximizing profit and constructing a society free from the burden of mutual responsibility—that is, a society whose essence is captured in the faces of children facing the terror of a future with no hope of survival.

# CHAPTER 2

# LOCKED UP: EDUCATION AND THE YOUTH CRIME COMPLEX

> *[I]n a time when punitive crime control measures have drastically increased, youth of color experience this hypercriminalization not only from criminal justice institutions but also from non–criminal justice structures traditionally intended to nurture: the school, the family, and the community center. Ultimately, in the era of mass incarceration, a "youth control complex" created by a network of racialized criminalization and punishment deployed from various institutions of control and socialization has formed to manage, control, and incapacitate black and Latino youth.*
>
> —Victor M. Rios, *"The Hypercriminalization of Black and Latino Male Youth in the Era of Mass Incarceration"*[1]

The voracious discourse and deformities of war as an organizer of collective experience took an ominous turn under the administration of George W. Bush. In response to the tragic attacks on September 11, the Bush administration not only made war and preemptive military strikes central to its foreign policy, but it also transformed the discourse of war into a regulatory principle for organizing everyday life. Against the threat of a terrorist attack, the Bush administration unleashed a Manichean imperative that short-circuited thought and gave free rein to the daily mobilization of mass-induced fear, rendering inessential the constitutive mechanisms of politics, particularly deliberative exchange based on reason and evidence, critical debate, shared responsibility, and ethical accountability. The discourse of the post-9/11 Bush administration was hypermasculine in tone and militaristic in response, legitimated in simplistic contrasts between good and evil.

Rather than invite deliberation and dialogue, abstract yet powerfully emotive language stifled thinking and squelched dissent. Bush-speak proved a profoundly antipolitical discourse, because it was incapable of imagining—and in fact disdained—a notion of politics based on judgment, meaningful criticism, and multiple public spheres.[2] As the rhetoric and terrain of politics were emptied of any democratic substance and war became the primary organizing principle of society, the state aligned itself more closely with corporate power, which it further strengthened by making corporations "self regulatory" even as it rather ironically bound its own citizens more tightly in a web of surveillance and control. Under this corporate model of politics, contemporary *raison d'etat* was no longer defined against economic, political, and social inequities: the state now restructured itself in the interests of finance capital, exercising authority through modes of governance that relied on fear, punishment, and the disciplinary organs of the punitive state.[3] This shift toward the corporate state, according to David Theo Goldberg, can be traced back to the mid-1970s when managing populations and markets became "central to the structural shifts in state formation away from welfarism and the caretaker state."[4] Over the past three decades, favorable memories of the liberal welfare state providing a social safety net for its citizens while improving the quality of their lives have been the object of unrelenting propaganda, excised from public memory and recalled with scorn for the "dependencies" it was alleged to have generated. Restoring the meaning and purpose of the social state, so viciously attacked by market fundamentalists, especially under the former Bush administration, is important, not only because it creates the conditions for democracy to become thinkable—and so possible—again. At stake are nothing less than the lives and futures of our children—all of them—especially those children most disadvantaged by market forces and a corporate state. Goldberg provides a synopsis of the social state that is worth repeating. He writes:

From the 1930s through the 1970s, the liberal democratic state had offered a more or less robust site of institutional apparatuses concerned in principle at least to advance the welfare of its citizens. This was the period of advancing social security, welfare safety nets, various forms of national health systems, the expansion of and investment in public education, including higher education, in some states to the exclusion of private and religiously sponsored educational institutions. It saw the emergence of state bureaucracies as major employers especially in later years of historically excluded groups. And all this, in turn, offered optimism among a growing proportion of the populace for access to middle-class amenities, including those previously racially excluded within the state and new immigrants from the global south.[5]

Under the emerging regime of neoliberalism, the modicum of social egalitarianism that was at the heart of the welfare state was both derided and dismantled, in spite of its partial successes. "Shock and awe," the military-inspired aesthetic used to launch the televised invasion of Iraq, was redeployed as a series of precise assaults on constitutional rights, dissent, and justice itself.[6] Torture, kidnapping, secret prisons, and the dissolution of habeas corpus—tacitly supported by a culture paralyzed by fear—became the protocol of the newly refashioned, repressive state, unapologetically engaged in illegal legalities abetted by a war culture that legitimated the expansion of state-sanctioned violence.[7] Rather than simply being weakened by the growing power of transnational corporations and the globalization of finance capital, the state was transformed from an already weakened welfare state into an increasingly powerful racialized warfare state.[8] As the war on terror was reconfigured and redeployed onto the domestic front, the mobilization of state violence required the recalibration of its racist logic, as those who increasingly became the objects of its power included people of color whose disposability was codified by their status as ghost detainees, administrative deportees, and enemy combatants, or in the curt label "collateral damage." The dark threat of totalitarian power was bolstered by the regressive impulse to view anyone considered a potential or actual terrorist as beyond the register of moral concern, undeserving of legal protections or moral rights.[9]

While the rise of the carceral state under the Bush administration has been the subject of intense debate in the last few years, what has been largely ignored is how the war at home both militarized public life and refashioned the criminal justice system, prisons, and even the schools, as preeminent spaces of racialized violence.[10] For many young people, the war at home has been transformed into a war against youth.[11] Historically, it has become commonplace for youth to be treated equivocally by adults as both a threat and a promise; the ambiguity that characterizes this mix of fear and hope has given way within the last 20 years to a much more one-sided and insidious view of young people as lazy, mindless, irresponsible, and even dangerous. Gone are the ideals, if not the utopian struggles, that promised young people a future that would exceed the limits and possibilities of the present. Dystopian fears about youth in the United States have intensified since the events of 9/11, as has the public's understanding of youth as an unruly and unpredictable threat to law and order. This tragedy is made obvious by the many "get tough" policies that render young people criminals and deprive them of basic health care and education, as state and federal funds for schools and child welfare services are cut back. Thus, the category of youth has been effectively eliminated from any discourse promoting the general welfare or the future of democracy. While the predicament of all youth under the regime of neoliberalism deepens in the midst of the current economic crisis, it does not affect all young people in the same way. More and more working-class and middle-class youth and poor youth of color either find themselves in a world with vastly diminishing opportunities or are fed into an ever-expanding system of disciplinary control that dehumanizes and criminalizes their behavior in multiple sites, extending from the home and school to the criminal justice system—not, of course, fed in order to be "absorbed" and "incorporated" into the system, but rather fed and vomited up, thus securing the permanence of their exclusion.

More and more youth have been defined and understood within a war on terror that provides an expansive, antidemocratic framework for referencing how they are represented, talked about, and inserted within a growing network of disciplinary relations that responds to the problems they face by criminalizing their behaviors and subjecting them to punitive modes of conduct. Youth in America have increasingly exhibited a series of disappearances, barely represented in humane terms in the public domain, and largely invisible in terms of their own needs. As the social state is reconfigured as a punishing state, youth become the enemy in hiding, dangerous bearers of unwanted memories. Progressively represented as troubling and a potential danger to society, they are scorned precisely because they offer a grim reminder of adult responsibility. Youth embody an ethical referent that *should* require adults to question the prevailing economic Darwinism and the future it emphatically denies in favor of an eternal present subject only to the market-driven laws of capital accumulation.

As the language of democracy is divested of concern for the future, adult obligations, and social responsibility in general, complex and productive representations of young people have gradually disappeared from public discourse only to reappear within the demonizing and punishing rhetoric of fear and crime. No longer inscribed in the metaphors of hope, youth—especially those marginalized by race and class—have now been cast into an ever-growing circle of groups targeted through the rhetoric of war and terrorism. Youth now occupy the status of what Bill Owens, the former conservative governor of Colorado, referred to as "a virus...let loose upon the culture."[12]

In an increasingly militarized society, the inventory of threats—inflected demographically through the taint of race and politically through the taint of socialism—have expanded to include not only immigrants, African Americans, Latinos, the government, high taxes, crime, godless sexual depravity, harassment, and acts of terror, but also youth in general and poor young minority males in particular. Fear, mistrust, and coercion are at the conceptual core of the war on terrorism. When these forces are aligned with the demonizing of youth by the media, scholars, politicians, and the general public, a lethal mix of hyperpunitive laws is produced that expands the circuits of repression and disposability designed to regulate the behavior of young people.[13] As Jean and John Comaroff have concluded, "the way young people are perceived, named, and represented betrays a lot about the social and political constitution of a society."[14]

Rendering poor minority youth as dangerous and a threat to society no longer requires allusions to biological inferiority; the invocation of cultural difference is enough to both racialize and demonize "difference without explicitly marking it"[15] in the post–civil rights era. This disparaging view of young people has promulgated the rise of a punishing and (in)security industry whose discourses, technologies, and practices have become visible across a wide range of spaces and institutions.[16] As the protocols of governance become indistinguishable from military operations and crime-control missions, youth are more and more losing the protections, rights, security, or compassion they deserve in a viable democracy. Rather than dream of a future bright with visions of hope, young people, especially youth marginalized by race and color, face a coming-of-age crisis marked by mass incarceration and criminalization, one that is likely to be intensified in the midst of the global financial, housing, and credit crisis spawned by neoliberal capitalism. Central to such a future is what Victor Rios calls a "youth control complex...an ecology of interlinked institutional arrangements that manages and controls the everyday lives of inner-city youth of color"; this complex has "a devastating grip on the lives of many impoverished male youth of color" and continues to promote the hypercriminalization of black and Latino youth.[17] One measure of this "youth control complex" is on full display in the state of Washington where 4th grade reading scores and graduation rates are used to determine how many prison cells will be built. As one teacher, Jesse Hagopian, points out, "So rest assured if your 9-year-old stumbles over syntax or has trouble sounding out the word 'priorities,' the state has readied the necessary cellblock accommodations."[18] Equally disconcerting is the lack of a public discourse—let alone outrage—capable of making a connection between this "youth control complex" and the broader structural forces that produce and sustain it. Angela Davis is instructive on this issue. She writes:

The incarceration of youth of color—and of increasing numbers of young women of color (women have constituted the fastest growing sector of the incarcerated population for some time now)—is *not* viewed as connected to the vast structural changes produced by deregulation, privatization, by the devaluation of the public good, and by the deterioration of community. Because there is no public vocabulary which allows us to place these developments within a historical context, individual deviancy is the overarching explanation for the grotesque rise in the numbers of people who are relegated to the country's and the world's prisons.[19]

With the election of Barack Obama, it has been argued that not only will the social state be renewed in the spirit and legacy of the New Deal but the punishing racial state will also come to an end.[20] From this perspective, Obama's election not only represents a post-racial victory but also signals a new space of post-racial harmony. In assessing the Obama victory, *Time* magazine columnist Joe Kline wrote: "It is a place where the primacy of racial identity—and this includes the old Jesse Jackson version of black racial identity—has been replaced by the celebration of pluralism, of cross-racial synergy."[21] Obama won the 2008 election because he was able to mobilize 95 percent of African Americans, two-thirds of all Latinos, and a large proportion of young people under the age of 30. At the same time, what is generally forgotten in the exuberance of this assessment is that the majority of white Americans voted for the John McCain/Sarah Palin ticket. While "post-racial" may mean less overt racism, the idea that we have moved into a post-racial period in American history is not merely premature—it is an act of willful denial and ignorance. Paul Ortiz puts it well in his comments on the myth of post-racialism:

The idea that we've moved to a post-racial period in American social history is undermined by an avalanche of recent events: the U.S. Supreme Court's dismantling of Brown vs. Board of Education and the resegregation of American schools; the Bush administration's response to Hurricane Katrina; the Clash of Civilizations thesis that promotes the idea of a War against Islam; the backlash facing immigrant workers and a grotesque prison industrial complex. [Moreover]...[w]hile Americans were being robbed blind and primed for yet another bailout of the banks and investment sectors, they were treated to new evidence from Fox News and poverty experts that the great moral threats facing the nation were greedy union workers, black single mothers, Latino gang bangers and illegal immigrants.[22]

Missing from the exuberant claims that Americans are now living in a post-racial society is the historical legacy of a neoconservative revolution, officially launched in 1980 with the election of Ronald Reagan, and its ensuing racialist attacks on the welfare "Queens"; Bill Clinton's cheerful compliance in signing bills that expanded the punishing industries; and George W. Bush's "willingness to make punishment his preferred response to social problems."[23] In the last 30 years, we have witnessed the emergence of policies that have amplified the power of the racial state and expanded its mechanisms of punishment and mass incarceration, the consequences of which are deeply racist—even as the state and its legal apparatuses insist on their own race neutrality. These racially exclusionary policies and institutions are not poised to disappear with the election of President Barack Obama.

The discourse of the post-racial state also ignores how political and economic institutions, with their circuits of repression and disposability and their technologies of punishment, connect and condemn many impoverished youth of color in the inner cities to persisting structures of racism that "serve to keep [them] in a state of inferiority and oppression."[24] Unfortunately, missing from the discourse of those who are arguing for the kind of progressive change the Obama administration should deliver is any mention of the crisis facing youth and the terrible toll it has taken on generations of poor white, black, and brown kids. Bringing this crisis to the forefront of the political and social agenda is crucial, particularly since Obama in a number of speeches prior to assuming the presidency refused to adopt the demonizing rhetoric often used by politicians when talking about youth. Instead, he pointedly called upon the American people to reclaim young people as an important symbol of the future and democracy itself:

[C]ome together and say, "Not this time." This time we want to talk about the crumbling schools that are stealing the future of black children and white children and Asian children and Hispanic children and Native American children. This time we want to reject the cynicism that tells us that these kids can't learn; that those kids who don't look like us are somebody else's problem. The children of America are not those kids, they are our kids.[25]

If Barack Obama's call to address the crucial problems facing young people in this country is to be taken seriously, the political, economic, and institutional conditions that both legitimate and sustain a shameful attack on youth have to be made visible, open to challenge, and transformed. This can happen only by refusing the somnambulance and social amnesia that coincide with the pretense of a post-racial politics and society, especially when the matter concerns young and poor people of color. To reclaim youth as part of a democratic imaginary and a crucial symbol of the future requires more than hope and a civics lesson: it necessitates transforming those power arrangements and market-driven values that have enabled the rise of the punishing state and have produced a polity that governs through the logic of crime and disposability—all the while disparaging the patriotism of critically engaged citizens who reject the role of either soldiers in the service of empire or consumers eager to boost the profits of corporate elites.

## DEPOLITICIZING THE SOCIAL AND PUNISHING YOUTH IN A SUSPECT SOCIETY

Under the regime of neoliberalism, a more ruthless form of economic Darwinism has emerged that assumes a position of moral neutrality—as allegedly mirrored in the abstract workings of the market—and undermines the bonds of the social by collapsing them into the realm of the private. Any notion of shared humanity and responsibility gives way to a survival-of-the-fittest mentality and fear for oneself—often coinciding with an indifference to the plight of others and to public considerations. As selfish market-driven interests increasingly trump social needs, scorn and contempt replace compassion for those bearing the burden of collectively induced misfortunes, such as poverty, unemployment, mortgage forfeitures, and other social ills. Under such circumstances, an ethic of cutthroat individual competition prevails, and the language of the social is either devalued or ignored. As part of a frontal assault on the institutions and values that make up the social state, neoliberal zealots define public goods as a form of pathology or deficit (as in public schools, public transportation, public welfare), while modeling all dreams of the future around the narcissistic, privatized, and self-indulgent needs of consumer culture and the dictates of the allegedly free market. Stripped of its ethical and political importance, the public sphere has been largely reduced to a space where private interests are displayed—and the social order increasingly mimics a giant reality TV show where any concept of the public is reduced to a conglomeration of private woes, tasks, conversations, and confessionals.

As the social state is hollowed out, the call for self-reliance replaces collective struggles for social justice, and the public's ability to translate private problems into both public concerns and collective action diminishes. As the social is devalued, public discourse and democratic politics disappear, only to be replaced by a litany of individual misfortunes to be borne in isolation. In the hyperindividualized society, principles of communal responsibility are undercut, derided, or erased; "individuals are called upon to invent and deploy individual solutions to socially produced discomforts," relying exclusively upon their own resources, skills, and wits.[26] Within this neoliberal moral economy, responsibility to oneself takes priority, and the ethical duty to care for others is diminished in value when those in need are not openly derided. Not surprisingly, under such circumstances, individual suffering no longer registers as a social concern, as all notions of injustice are assumed to be the outcome of personal failings or deficits. Signs of this pathologizing of marginalized individuals and the social sphere as a whole can be found everywhere. Poverty is now imagined to be a problem of individual character. Racism is now understood as merely an act of individual discrimination (if not discretion), and homelessness is reduced to a choice made by lazy people. Not only has the concept of the social largely faded out of view during the last three decades, but politics itself was now mediated through a pervasive spectacle of terrorism in which fear and violence became the only modalities through which to grasp the meaning of the self and larger social relations.

As the modernist dream of infinite progress for each succeeding generation erodes even further under the current global meltdown, minority youth are increasingly excluded from decent jobs, health care, and social services, while being even more insistently subject to the terrors of the present economic crisis. And just as major problems such as racism, homelessness, and persistent poverty disappear from the inventory of public considerations, social investments are replaced by penal solutions, giving rise to a punishing state that removes from the social order those who have no market value, those who are fatally defined as flawed consumers, and those who are designated "other" through an often-groundless association with crime, redundancy, poverty, or simply disposability. As Zygmunt Bauman puts it,

Youth are now recast as collateral casualties of consumerism, the poor are now and for the first time in recorded history purely and simply a worry and a nuisance. . . . They have nothing to offer in exchange for the taxpayers' outlays. . . . While the poor are banished from the streets, they can also be banished from the recognizably human community: from the world of ethical duties. This is done by rewriting their stories away from the language of deprivation to that of depravity. The poor are portrayed as lax, sinful, and devoid of moral standards. The media cheerfully cooperate with the police in presenting to the sensation-greedy public lurid pictures of the "criminal elements," infested by crime, drugs and sexual promiscuity, who seek shelter in the darkness of their forbidding haunts and mean streets. The poor provide the usual suspects to be round up, to the accompaniment of a public hue and cry, whenever a fault in the habitual order is detected and publicly disclosed.[27]

The increasing privatization of public interests and the moral hardening of the social order, largely shaped by the biopolitical project of neoliberalism, have undermined the ethical and political fabric of public life. The result is the production of new strategies of governance, largely mediated through a combination of fear, the politics of (in)security, and the criminalization of social problems, leading to the spread of values, policies, practices, and technologies of the punishing state to public spheres traditionally removed from such influences.

While the rise of neoliberalism has undermined the most basic values and institutions of democracy in the United States, it has had a particularly devastating effect on youth, as the combined modalities of regulation, control, surveillance, and punishment radically alter the public spheres inhabited by minority youth. While all youth are now suspect, poor minority youth have become especially targeted by modes of social regulation, crime control, and disposability that have become the major prisms that now define many of the public institutions and spheres that govern their lives.[28] The model of policing that now governs all kinds of social behaviors constructs a narrow range of meaning through which young people define themselves. This rhetoric and practice of policing, surveillance, and punishment have little to do with the project of social investment and a great deal to do with increasingly powerful modes of biopolitical regulation, pacification, and control—together comprising a "youth control complex" whose prominence in American society points to a state of affairs in which democracy has lost its claim and the claiming of democracy goes unheard. The United States' claim to democracy, already weakened on a global level by the go-it-alone attitude that precipitated the war in Iraq, loses much more of its credibility as a democratic nation when one considers the degree to which militarized relations of war within its own borders now constitute how minority youth are understood and treated by much of adult society. The military character of the war waged against young people is best exemplified by the ascendancy of the prison as a definitive model of disciplinary regulation and a primary element of governance in dealing with disposable populations on the domestic front.

The prison symbolizes not merely the failure of social reform and the emerging politics of a racially predicated logic of disposability, but also a prominent element in the war against poor youth who are no longer considered fit to be soldiers, consumers, or advertising billboards for corporate profits. Instead, they are viewed as an excess, a cancer on the body politic that must be removed to protect the safety and health of the larger society. Under such circumstances, the prison takes on a new purpose and meaning in American society, one that grants an afterlife to an authoritarianism that pushes beyond the boundaries of legitimate governmental practice.[29] Angela Davis extends this argument, insisting that the prison is the institution par excellence in the aftermath of the breakdown of the welfare state:

[The prison in] U.S. society has evolved into that of a default solution to the major social problems of our times.... [I]mprisonment is the punitive solution to a whole range of social problems that are not being addressed by those social institutions that might help people lead better, more satisfying lives. This is the logic of what has been called the imprisonment binge: Instead of building housing, throw the homeless in prison. Instead of developing the educational system, throw the illiterate in prison. Throw people in prison who lose jobs as the result of de-industrialization, globalization of capital, and the dismantling of the welfare state. Get rid of all of them. Remove these dispensable populations from society. According to this logic the prison becomes a way of disappearing people in the false hope of disappearing the underlying social problems they represent.[30]

The centrality of the prison as a disciplinary, regulatory, and pedagogical model suggests that the carceral apparatuses of the twenty-first century may emerge in a distinctive and perhaps even more ruthless form than its predecessors, particularly as strategies of governance and modes of sovereign power increasingly mirror the savage brutalities of the market.[31] In its language, practices, and policies, neoliberalism not only "extends the rationality of the market [into] domains that are not primarily economic"[32] but also creates more punishing modes of governance. This is a mode of biopolitics that renders market interests invisible by insisting that its primary goal is to promote the security and welfare of a human life: an unregulated market is the best caretaker of people's needs. In actuality, its real purpose is to collapse the distinctions between crime and social problems, prison and school, and race and disposability, while constructing spaces that subject minority youth and others rendered redundant to a form of punitive control, if not social death. Punishment and incarceration, long absolved of the pretense of rehabilitation, are now primarily contained within what Zygmunt Bauman has called "the human waste disposal industry."[33] At its center is a network of institutions "obsessed with surveillance, security, and punitive penal practices"[34] that not only reproduces racial inequality, social wretchedness, and individual suffering but also "serve[s] as a main socializing and controlling agent for black and Latino youth who have been labeled 'deviant.'"[35] There is more at stake here than a politics of fear, discipline, and control: a mode of governance is emerging that deprives many young people of a childhood and forecloses for them the possibility of a meaningful future.

## GOVERNANCE, CRIME, AND THE PRISON-CONTROL COMPLEX

In the 1970s, as the regime of neoliberalism and the rationality of the market gradually came to dominate most aspects of American life, the war against the legacy of the New Deal and against the cultural revolution of the previous decade took on a new dimension. The political realm shifted from understanding the difficulties facing individuals within the context of surrounding structural constraints and socially inscribed forms of injustice to attributing personal responsibility to the individuals themselves. In conjunction with the dismantling of most remnants of the welfare state, the state intensified its more repressive modes of power and increasingly relied on appeals to fear to usher in a kind of politics in which the modalities of crime and punishment exercised a powerful influence on how Americans viewed themselves and their relations to others and the larger social order.[36] One consequence was that the war against poverty was replaced by the war against crime, just as the welfare state and its support for a social safety net were replaced by a punishing state and its call for criminalizing behaviors generally associated with the structured inequities of the social order. In addition, the shift to governing through the lens of crime and fear also inspired a massive redistribution of resources away from the welfare state to the punishing state. David Theo Goldberg's reflections on the transformation from the welfare state to the repressive neoliberal state are revealing and serve as a backdrop to the war against youth and the rise of a mode of governance through crime, exclusion, and disposability. Contrary to advocates of neoliberalism who claim their policies minimize "big government," Goldberg argues that the neoliberal state now exerts more power and control:

Where the prevailing social commitments for the liberal democratic state had to do with social well-being revealed in the registers of education, work, health care, and housing, the neoliberal state is concerned above all with issues of crime and corruption, controlling immigration and tax-cut-stimulated consumption, social control and securitization. So the contemporary slogan of neoliberalism might as well be "The state looks after your interests by encouraging you to choose to lock yourself into gated communities (while it locks up the undesirable in prisons) or locks out the externally threatening (by way of immigrations restrictions)." Where the liberal democratic state was concerned in the final analysis with the welfare of its citizens, all the contradictions of its arrangement and application notwithstanding, the neoliberal state is concerned above all with their security. The "social security" state has morphed in meaning from prevailingly economic significance to its more assertively disciplinary interventions. If the social welfare state could be seen as modestly paternalistic, the neoliberal state has proved invasively repressive.[37]

Jonathan Simon has argued that since the 1980s, the public's desire for safety and its fear of crime have provided the impetus for "a new civil and political order structured around the problem of violent crime."[38] According to Simon, crime has not only become central to how authority is exercised in the United States but has also ushered in a new mode of politics that merges "the penal state and the security state."[39] For Simon, the discourse of crime and punishment has become both an axis for how Americans come to "know and act on ourselves, our families, and our communities" and a structuring principle for reworking how various institutions are perceived and organized under a repressive state apparatus.[40] As a pervasive and relentless war on terror elevated all citizens to the status of potential enemies of the state, new technologies of surveillance and control spread throughout the social order, while the practices of punishing, repressing, and exercising state power over people took on a new urgency as a matter of governance. Simon argues that this aggressive rhetoric of crime and punishment constitutes not only a crisis of politics but also a crisis of governance, one that he labels the new politics of "governing through crime." One consequence of governing through crime has been the development of "the imprisonment binge" of the last 30 years.[41] While the imprisonment binge of the last few decades is central to this emphasis on crime, it has taken on a new importance and influence as crime has now become one of the major organizing principles through which "other problems are recognized, defined, and acted upon—and social relations constructed."[42] In addition, crime now becomes an excuse not only to expand modes of security, surveillance, and control throughout society, but also to retool the inheritances of racism through a mode of governance that takes as one of its objectives the punishing, if not removal from the social body, of poor black and brown youth who are viewed as excess and rendered disposable. What is important about Simon's concept of governing through crime is its recognition of the emergence of a more capacious model of criminalization that reached its apex under the George W. Bush administration, one that makes crime "the central tool for governing the everyday citizen, even if he or she has never committed a crime. Crime and punishment have been prioritized in the United States to influence the actions of the everyday citizen."[43] In what follows, I want to focus on how governing through crime and the politics of disposability have helped to shape the cultural politics of an economy of punishment and its devastating effects on poor black and brown youth in the United States.

At the center of the politics of social control through criminalization are a series of social relationships in which the prison has become a model to solve a wide range of social, economic, and political problems.[44] At the core of this approach is a steadfast principle of what can be called a racialized economic Darwinism, one that is central to the prevailing neoliberal logic of the free market. Under such conditions, as Zygmunt Bauman insists, collectively caused problems are now interpreted as "an individually committed sin or crime. . . . Prisons now deputize for the phased-out and fading welfare institutions, and in all probability will have to go on readjusting to the performance of this new function as welfare provisions continue to be thinned out."[45] As the politics of the social state gives way to the biopolitics of disposability, the prison becomes a preeminently valued institution whose disciplinary practices become a model for dealing with the increasing number of young people who are considered to be the waste products of a market-mediated society. As Simon points out, what is unique about the contemporary prison is that it unapologetically now functions as a warehouse and waste-disposal factory. He writes:

The distinctive new form and function of the prison today is a space of pure custody, a human warehouse or even a kind of social waste management facility, where adults and some juveniles distinctive only for their dangerousness to society are concentrated for purposes of protecting the wider community. The waste management prison promises no transformation of the prisoner through penitence, discipline, intimidation, or therapy. Instead, it promises to promote security in the community simply by creating a space physically separated from the community in which to hold people whose propensity for crime makes them appear an intolerable risk for society.[46]

The institution of the prison is at the ideological center of the biopolitics of the punishing state dutifully inscribing its presence into the political and cultural landscape of everyday life. As Angela Davis reminds us, what is important to recognize is that the prison-industrial complex now embraces a vast set of institutions that constitute the disciplinary apparatuses at the heart of the punishing state. According to Davis, the network of institutions

includes state and federal prisons, county jails, jails in Indian country, detention centers run by the Department of Homeland Security, territorial prisons in areas the U.S. refuses to acknowledge as its colonies, and military prisons—both within the U.S. and outside of its borders. The population growth in domestic prisons, the emergence of new industries dependent on this growth, the retooling of old industries to accommodate and profit from imprisonment, the expansion of immigrant detention centers, and the use of military prisons as a major weapon in the so-called war on terror, the articulation of anti-crime rhetoric with anti-terrorism rhetoric—these are some of the new features of the prison-industrial-complex.[47]

The institution of the prison symbolizes the power of the repressive state operating under the guise of the war on terror, while its growing presence and influence normalizes a racially predicated politics of disposability. Moreover, it extends its core values, modes of discipline, and parameters of control to a vast array of other institutions outside of the prison-industrial complex, creating what Ruth Wilson Gilmore calls a "tale of fractured collectivities—economies, governments, cities, communities, and households."[48] As crime, imprisonment, and punishment become central features of the punishing state, the policies and practices of governing through crime are no longer limited to urban centers of deep poverty and social dislocation but now spread to those locations and "spatial sites where middle-class life is performed on an everyday basis: office buildings, universities, daycare centers, medical complexes, apartment buildings, factories, and airports."[49] The hard logic and raw impact of governing through a culture of fear and the power to punish can be grasped, in part, by the degree to which imprisonment, punishment, and detention have become both the preferred responses to social problems and a linchpin of the new political order and its disciplinary mode of punitive governance.

As the culture of control, punishment, and disposability become a central force in shaping the fabric of American life, it has found expression in policies and legislation at all levels of government that give more power to prosecutors and the police, while limiting the discretionary power of judges and the courts. Calls for the death penalty and harsher laws such as "three strikes" measures and other sentencing enhancements coupled with the demand for more prisons now dominate the rhetoric of politicians playing to media-induced moral panics about crime while undermining the possibilities of democratic modes of governance and justice. As governance is increasingly predicated on war as the primary logic for shaping daily life,[50] the ever-growing prison-industrial complex and its project of mass imprisonment have taken on a particularly toxic register. Since the 1970s, under the repressive state's biopolitical commitment to neoliberalism, building prisons has become America's housing policy for the poor, signaling an attack not only on those for whom class and race loom large, but also on a generation of young people who have few rights, and even less power, and have come to symbolize a drain on potential profits (given the cost of providing them with even a minimal level of quality education, health care, employment, housing, and income). What are we to make of the following shifts in carceral practices? According to a recent report released by the Pew Public Safety Performance Project,

Three decades of growth in America's prison population [have] quietly nudged the nation across a sobering threshold: for the first time, more than one in every 100 adults is now confined in an American jail or prison.... The United States incarcerates more people than any country in the world, including the far more populous nation of China. At the start of the new year, the American penal system held more than 2.3 million adults. China was second, with 1.5 million people behind bars, and Russia was a distant third with 890,000 inmates, according to the latest available figures. Beyond the sheer number of inmates, America also is the global leader in the rate at which it incarcerates its citizenry, outpacing nations like South Africa and Iran.[51]

As shocking as these figures are, they are particularly grave for people of color and reveal how the punishing state invests in the prison-industrial complex as a way of managing large populations of people of color who have been rendered disposable, shorn of their rights, and deemed unfit for state protection. As Angela Davis points out, "In 1985, there were fewer than 800,000 people behind bars. Today there are almost three times as many imprisoned people and the vast increase has been driven almost entirely by the practices of incarcerating young people of color."[52] For instance, one in 36 Hispanic adults is behind bars, while "one in every 15 black males aged 18 or older is in prison or jail."[53] In fact, young black men between the ages of 20 and 34 are jailed at a rate of one in nine. Moreover, a full 60 percent of black high school dropouts, by the time they reach their mid-thirties, will be prisoners or ex-cons.[54] This apartheid-based system of incarceration bodes especially ill for young black males. According to Paul Street:

It is worth noting that half of the nation's black male high school dropouts will be incarcerated—moving, often enough, from quasi-carceral lock-down high schools to the real "lock down" thing—at some point in their lives. These dropouts are overrepresented among the one in three African American males aged 16 to 20 years old who are under one form of supervision by the U.S. criminal justice system: parole, probation, jail, or prison.[55]

As Loic Wacquant points out, racially targeted "get tough" crime policies produce their counterpart in racially skewed forms of mass imprisonment, legitimized, in part, by "the reigning public image of the criminal" as

that of a *black* monster, as young African American men from the "inner city" have come to personify the explosive mix of moral degeneracy and mayhem. The conflation of blackness and crime in collective representation and government policy (the other side of this equation being the conflation of blackness and welfare) thus re-activates "race" by giving a legitimate outlet to the expression of anti-black animus in the form of the public vituperation of criminals and prisoners.[56]

Moreover, such policies both sanction and promote race-based drug arrests for drug sales and possession, filling prisons with young black men "who are nearly twelve times as likely to be imprisoned for drug convictions as adult white men," while promoting vast racial disparities in the nation's prisons.[57]

The frontier mentality shaping punishment and mass imprisonment exacts a heavy price on impoverished youth of color, while eviscerating institutions designed to benefit the public good. The financial costs alone of maintaining this prison culture are extravagant, blowing a massive hole through tattered state budgets while undermining their most basic public services, including education and health care. And these trends will become more exacerbated as tax revenues decline and social services are stretched to the limits under the strain of the current financial and credit crisis. In 2008, "31 states had budget gaps totaling $40 billion";[58] consequently, many states had to slash school financing, decrease the number of subsidized meals available for poor children, and reduce, in some cases, the number of days children attend school. Sadly, the situation will get worse before it gets better.

In the face of these cutbacks, states will continue to dispense huge amounts of money into a bloated and overextended prison system. According to the Pew report,

In 1987, the states collectively spent $10.6 billion of their general funds—their primary pool of discretionary tax dollars—on corrections. [In 2007], they spent more than $44 billion, a 315 percent jump, data from the National Association of State Budget Officers show. Adjusted to 2007 dollars, the increase was 127 percent. Over the same period, adjusted spending on higher education rose just 21 percent.... Total state spending on corrections—including bonds and federal contributions—topped $49 billion last year, up from $12 billion in 1987. By 2011, continued prison growth is expected to cost states an additional $25 billion.[59]

A more recent Pew Center study reports, "For all levels of government, total corrections spending has reached an estimated $68 billion, and increase of 330 percent since 1986.... Only Medicaid spending grew faster than spending on corrections."[60] Even as violent crime fell by 25 percent in the past 20 years, states increased their spending on corrections, with 13 states now spending more than one billion dollars a year in general funds on their corrections systems. Many states are spending more on corrections than they are on higher education, while jettisoning a range of important social programs that provide for people's welfare.[61] For example, James Sterngold reported in 2007 that "[b]ased on current spending trends, California's prison budget will overtake spending on the state's universities in five years."[62] In this particular instance, the shift to governing through crime makes a mockery of a state that lays claim to smart policymaking. While the average cost to imprison someone is $23,876, some states such as Rhode Island pay out as much as $45,000 per inmate. A number of states because of the economic recession are passing legislation to reduce prison sentences and the cost of the imprisonment binge. In many states, it costs far more to imprison people than it does to provide them with a decent education. What is so tragic about these figures is that 50 percent of the people who are behind bars are there for nonviolent crimes, while 70 percent of all inmates are people of color. Clearly, there is more at work here than a prison-industrial complex that amounts to the squandering of human and financial resources at massive taxpayer expense: what Loic Wacquant rightly calls "a de facto policy of carceral affirmative action towards African Americans"[63] is operating to produce largely ignored collateral effects that extend the impact of the punishment industry far beyond the walls of prison culture.

As Jason DeParle has argued, mass incarceration of poor men and youth of color "deepens the divides of race and class" by "walling off the disadvantaged, especially unskilled black men, from the promise of American life."[64] Imprisonment makes black inmates poorer because they are not given the opportunity to learn a skill or get an education, and when they are released their chances of getting a job are slim, especially in an economy in which the rate of unemployment among blacks is twice that of whites. Though education is typically a prerequisite for employment, many former convicts are excluded from various forms of student aid because of certain previous crimes such as a drug conviction. Ex-prisoners are also excluded from even a modicum of social provision and income by being denied welfare payments, Medicaid, veterans' benefits, food stamps, and in some cases public housing. Under such exclusionary practices, African American males suffer a number of indignities and restrictions on their rights that prevent them from integrating into mainstream society. Brent Staples provides a further snapshot of some of the inhuman forces at work in placing sanctions on black men once they enter the criminal justice system. He writes:

Ex-cons are marooned in the poor inner-city neighborhoods where legitimate jobs do not exist and the enterprises that led them to prison in the first place are ever present. These men and women are further cut off from the mainstream by sanctions that are largely invisible to those of us who have never been to prison. They are commonly denied the right to vote, parental rights, drivers' licenses, student loans and residency in public housing—the only housing that marginal, jobless people can afford. The most severe sanctions are reserved for former drug offenders, who have been treated worse than murderers since the start of the so-called war on drugs. The Welfare Reform Act of 1996, for example, imposed a lifetime ban on food stamps and welfare eligibility for people convicted of even a single drug felony.[65]

The racially defined nature of the punishing state is also evident in the grim facts that "the average state disenfranchises 2.4 percent of its voting-age population—but 8.4 percent of its voting-age blacks. In fourteen states, the share of blacks stripped of the vote exceeds 10 percent. And in five states (including Kentucky), it exceeds 20 percent. Focusing on black men . . . felony laws keep nearly one in seven from voting nationwide."[66] The racialized aftereffects of the punishing state's prison culture are also evident in the shattered families and fatherless children that populate many of America's impoverished cities. DeParle argues that "[f]rom 1980 to 2000, the number of children with fathers behind bars rose sixfold to 2.1 million.

Among white kids, just over 1 percent have incarcerated fathers, while among black children the figure approaches 10 percent."[67] Racially skewed crime policies do little to serve and protect the public, but can generate enormous profits for rich investors. As more and more prisons become privatized, the connection between mass incarceration and economic Darwinism takes on a foreboding register. Under the biopolitics of neoliberalism, criminalization produces black and brown bodies for a prison industry that pays high dividends to shareholders, promotes the growth of powerful prison guard unions, and attracts the support of varied special interests who view the prison-industrial complex as a low-risk investment with windfall profits. These special interests have become so powerful that they organized in 2008 to defeat Prop 5, a ballot initiative that sought to tackle many of the chronic problems facing a criminal justice system in California that is both deeply flawed and dysfunctional. In this case, prominent politicians across party lines joined with corporate interests and the powerful California prison guards' union, which provided $1.8 million to the campaign, to defeat the measure. Not only would the bill have reduced prison overcrowding, enhanced public safety, decreased costs, expanded drug treatment programs inside state prisons, and started the first drug treatment program for at-risk youth,[68] it is estimated that Prop 5 would have saved California taxpayers "at least $2.5 billion, according to the state's Legislative Analyst."[69]

The marriage of economic Darwinism and the racialized punishing state is also on full display in East Carroll Parish in Louisiana, where inmates provide cheap or free labor at barbecues, funerals, service stations, and a host of other sites. According to Adam Nossiter, "the men of orange are everywhere" and people living in this Louisiana county "say they could not get by without their inmates, who make up more than 10 percent of its population and most of its labor force. They are dirt-cheap, sometimes free, always compliant, ever-ready and disposable.... You just call up the sheriff, and presto, inmates are headed your way. 'They bring me warm bodies, 10 warm bodies in the morning,' said Grady Brown, owner of the Panola Pepper Corporation. 'They do anything you ask them to do.'... 'You call them up, they drop them off, and they pick them up in the afternoon,' said Paul Chapple, owner of a service station."[70] Nossiter claims that the system is jokingly referred to by many people who use it as "rent a convict" and is, to say the least, an "odd vestige of the abusive convict-lease system that began in the South around Reconstruction."[71]

Treating prisoners as commodities to be bought and sold like expendable goods suggests the degree to which the punishing state has divested itself of any moral responsibility with regard to those human beings who in the market-driven logic of neoliberalism are considered either commodities or disposable waste products. At the same time, as the beginning of an era of post-racialism is celebrated and racism is presumed to be an anachronistic vestige of the past in light of Barack Obama's election to the presidency, the workings of the punishing state are whitewashed and differentiated from racialized violence as the governing-through-crime complex is rendered invisible. Consequently, the American public becomes increasingly indifferent to the ways in which neoliberal rationality—with its practices of market deregulation, privatization, the hollowing out of the social state, and the disparaging of the public good—wages a devastating assault on African American and Latino communities, young people, and increasingly immigrants and other people of color who are relegated to the borders of American normalcy and patriotism. The punishing state not only produces vast amounts of inequality, suffering, and racism, but also propagates collective amnesia, cynicism, and moral indifference. Hence, there are few attempts in the dominant media to connect the problems in the prison system, particularly its deeply entrenched structures of racism, to the related crises of governance and the politics of the youth crime complex.

## YOUTH AND THE POLITICS OF PRISON CULTURE

As the punishing state gains in power, the prison-industrial complex is nurtured and supported by broader economic, political, and social conditions; its deeply structured racist principles, politics of disposability, and modes of authoritarian governance become part of the fabric of common sense, an unquestioned element of effective governance. As a disciplinary model, the prison reinforces modes of violence and control that are now central to the efforts of the punishing state to align its values and practices with a number of other important commanding social institutions. The reach of prison culture and its punitive disciplinary practices now extend into the home, workplace, juvenile criminal services, the school, and the entertainment industry. Along with growing incarceration rates for youth of color, young people now have to endure drug tests, surveillance cameras, invasive monitoring, home visits by probation officers, security forces in schools, and a host of other militarizing and monitoring practices used to target potential criminals, terrorists, and other groups represented as a threat to the state. Of course, under the Bush administration those who disagreed with the administration's domestic and foreign policy goals or whose skin color was dark were with a few exceptions regarded as a high security risk and as potential terrorists.[72] Unfortunately, as the arm of prison culture continues to spread throughout the society, it increasingly reinforces and provides a model for other institutions that deeply influence the lives of young people, exacting a terrible toll especially on the lives and futures of poor black and brown youth.

As traditional supports and social safety nets provided by the liberal social contract disappear, the condition of American youth deteriorates most visibly in the way in which they are stereotyped, demonized, and removed from the register of social concerns. With the rise of a mode of governance mediated through an emphasis on crime and the politics of disposability, youth become the new targets of a suspect society. As the ideologies and disciplinary practices of prison culture are incorporated into the pedagogies of the school and the criminal justice system—celebrated in various modes of mainstream entertainment—youth are increasingly subject to policies and practices suggesting they are worthy of no other treatment than that accorded to criminals—and this judgment is rendered without the benefit of trial, or the presumption of innocence.

Social violence evokes a special kind of cruelty when applied to children, and yet it has gained widespread support both in the public mind and in the deeply rooted rituals of popular culture that thrive on an ideology of masculine hardness, humiliation, and violence, rendering its participants indifferent to the suffering of others. Zygmunt Bauman has argued that "[e]very [society] produces its own visions of the dangers that threaten its identity, visions made to the measure of the kind of social order it struggles to achieve or to retain.... [T]hreats are projections of a society's own inner ambivalence, and anxieties born of that ambivalence, about its own ways and means, about the fashion in which that society lives and intends to live."[73] As a symbol of ambivalence, rather than a social investment or a population in need of protection and support, youth are now perceived as a threat to the crumbling social order. One response to this perceived danger is the emergence of a neoliberal state that seeks to bolster its weakened sovereignty by recasting youth as a threat to society and to gain its legitimacy by dealing with that threat—or being seen to deal with it accordingly (typically through media spectacles). As Lawrence Grossberg puts it,

Over the past twenty-five years, there has been a significant transformation in the ways we talk and think about kids and, consequently, in the ways we treat them. We live, for at least part of the time, in a rhetorically constructed picture of kids out of control, an enemy hiding within our most intimate spaces. The responses—zero tolerance, criminalization and imprisonment, psychotropic drugs and psychiatric confinement—suggest not only that we have abandoned the current generation of kids but that we think of them as a threat that has to be contained, punished, and only in some instances, recruited to our side.... [E]very second, a public high school student is suspended; every ten seconds, a public school student is corporally punished; every twenty seconds, a kid is arrested. Criminalization and medicalization are cheap (financially and emotionally) and expedient ways to deal with our fears and frustrations.[74]

When youth occupy the larger screen culture, they are represented mostly through images that are degrading and demonizing. It is difficult to find in the dominant media any sympathetic representations of young people who experience difficult times as a result of the economic downturn, the simultaneous erosion of security (around health care, work, education), and the militarization of everyday life. Youth are no longer categorized as Generation X, Y, and Z. On the contrary, they are now defined rhetorically in mainstream corporate media as "Generation Kill" or "Killer Children."[75] In the aftermath of the shooting rampages at Columbine High School and Virginia Tech, kids are largely defined through the world of frenzied media spectacles driven by sensationalist narratives and youth panics. Rather than being portrayed as victims of a "crisis of masculinity and male rage, an out-of-control gun culture, and a media that projects normative images of violent masculinity and makes celebrities out of murderers,"[76] youth are represented as psychologically unhinged, potentially indiscriminate killers (especially young returning veterans), gang rapists (falsely accused Duke University lacrosse players), school shooters, and desensitized domestic terrorists. Newspapers and other popular media offer an endless stream of alarming images and dehumanizing stories from the domestic war zone, allegedly created by rampaging young people. One typical newspaper account described how a group of third graders in south Georgia brought a knife, duct tape, and handcuffs to school as part of a plan to attack their teacher.[77] CNN's Anderson Cooper hosted a special report on school shootings on April 27, 2007, with the title "Killers in Our Midst," which not only capitalized on shocking and sensational imagery that swelled the network's bottom line but also added fuel to a youth panic that insidiously portrays young people as pint-size nihilists, an ever-present threat to public order.

Scapegoated youth thus provide the means for turning public attention away from alarming instances of state violence against thousands of detainees held in various secret prisons around the world, the outsourcing of torture by the CIA to Syria and other authoritarian regimes, the illegal legalities of an imperial presidency including the world-record-shattering incarceration rates of people of color in jails or prisons, and the endless abuses that young people suffer at the hands of adults in a geography of heightened poverty, racism, unemployment, and inequality.[78] And yet, while the public is flooded with reports of feral teenage boys poised to commit brutal, remorseless crimes, reinforcing the new common sense that the categories of "youth" and "super-predator" are synonymous, we hear little from the dominant media about either shocking rates of youth poverty and homelessness, or the 4 million youth "who are not in school and basically have no hope of finding work."[79] Nor is there the slightest public concern about the sharp rise over the last decade in the use of potent antipsychotic prescription drugs, stimulants, and antidepressants to medicate children and adolescents for a multitude of heretofore normal "teen" behaviors, ranging from mood swings to "oppositional defiant disorder."[80] Nor does the public hear much about the fate of young people in unregulated so-called "therapeutic schools whose 'tough love' treatments include having a bag placed over their head and a noose around their neck."[81] As Alex Koroknay-Palicz argues, "Powerful national forces such as the media, politicians and the medical community perpetuate the idea of youth as an inferior class of people responsible for society's ills and deserving of harsh penalties."[82] While such negative and demeaning views have had disastrous consequences for young people, under the reign of a punishing society and the deep structural racism of the criminal justice system, the situation for a growing number of young people and youth of color is getting much worse.

The suffering and deprivation experienced by millions of children in the United States in 2009—bound to become worse in the midst of the current economic meltdown—not only testifies to a state of emergency and a burgeoning crisis regarding the health and welfare of many children, but also bears witness to—and indeed indicts—a model of market sovereignty and a mode of punitive governance that have failed both children and the promise of a substantive democracy. The Children's Defense Fund in its 2008 annual report offers a range of statistics that provide a despairing glimpse of the current crisis facing too many children in America. What is one to make of a society marked by the following conditions:

- Almost 1 in 13 children in the United States live in poverty—5.8 million in extreme poverty.
- One in 6 children in America is poor. Black and Latino children are about 3 times as likely to be poor as white children.

- 4.2 million children under the age of 5 live in poverty.
- 35.3 percent of black children, 28.0 percent of Latino children, and 10.8 percent of white, non-Latino children live in poverty.
- There are 8.9 million uninsured children in America.
- One in 5 Latino children and 1 in 8 black children are uninsured, compared to 1 in 13 white children.
- Only 11 percent of black, 15 percent of Latino, and 41 percent of white eighth graders perform at grade level in math.
- Each year 800,000 children spend time in foster care.
- On any given night, 200,000 children are homeless, one out of every four of the homeless population.
- Every 36 seconds a child is abused or neglected, almost 900,000 children each year.
- Black males ages 15 to 19 are about eight times more likely to be gun homicide victims than white males.
- Although they represent 39 percent of the U.S. juvenile population, minority youth represent 60 percent of committed juveniles.
- A black boy born in 2001 has a one in three chance of going to prison in his lifetime; a Latino boy has a one in six chance.
- Black juveniles are about four times as likely as their white peers to be incarcerated. Black youths are almost five times as likely and Latino youths about twice as likely to be incarcerated as white youths for drug offenses.[83]

These figures suggest that young people in the United States are increasingly being constructed in relation to a future devoid of any hope. The notion that children should be treated as a crucial social resource and represent for any healthy society important ethical and political considerations about the quality of public life, the allocation of social provisions, and the role of the state as a guardian of public interests appears to be lost. The visual geographies and ever-expanding landscapes of violence young people inhabit provoke neither action nor ethical discrimination on the part of adult society, which might serve to prevent children from being relegated to our lowest national priority in the richest country in the world.

If prison is the ultimate expression of social exclusion for adults in the United States, managing and regulating youth through the lens of crime and repression represents its symbiotic underside. One consequence is that the most crucial institutions affecting the lives of young people are now under the influence of disciplinary apparatuses of control and repression that have become the most visible indicator of the degree to which the protected space of childhood, if not democracy itself, is being destroyed. As minority youth are removed from the inventory of ethical and political concerns, they are treated as surplus populations, assigned to a form of social death. In a suspect society that governs through a ruthless economic Darwinism, a sensationalized culture of violence, and the topology of crime, youth become collateral damage, while democratic governance disappears along with the moral and political responsibilities necessary for creating a better and more just future for succeeding generations.

Under the reign of a punishing mode of sovereignty, a racialized criminal justice system, and a financial meltdown that is crippling the nation, the economic, political, and educational situation for a growing number of young people and youth of color has gone from bad to worse. As families are being forced out of their homes because of record-high mortgage foreclosures and many businesses declare bankruptcy, tax revenues are declining and effecting cutbacks in state budgets, further weakening public schools and social services. The results in human suffering are tragic and can be measured in the growing ranks of poor and homeless students, the gutting of state social services, and the sharp drop in employment opportunities for teens and young people in their twenties.[84] Within these grave economic conditions, children disappear, often into bad schools, prisons, foster care, and even into their graves. Under the biopolitics of neoliberalism, the punishing state has no vocabulary or stake in the future of poor minority youth, and increasingly in youth in general. Instead of being viewed as impoverished, minority youth are seen as lazy and shiftless; instead of being recognized as badly served by failing schools, they are labeled uneducable and pushed out of schools; instead of being provided with decent work skills and jobs, they are either sent to prison or conscripted to fight in wars abroad; instead of being given decent health care and a place to live, they are placed in foster care or pushed into the swelling ranks of the homeless. Instead of addressing the very real dangers that young people face, the punishing society treats them as suspects and disposable populations, subjecting them to disciplinary practices that close down any hope they might have for a decent future. Perhaps the most powerful site in which these disciplinary practices are at work and bear down daily on the lives of many young people, but especially on the lives of minority youth, is in U.S. public schools, which now prepare many students for entry *not* into universities or colleges but into the juvenile criminal justice system.

## MILITARIZING PUBLIC SCHOOLS

The shift to a society now governed through crime, market-driven values, and the politics of disposability has radically transformed the public school as a site for a civic and critical education. One major effect can be seen in the increasingly popular practice of organizing schools through disciplinary practices that closely resemble the culture of the prisons.[85] For instance, many public schools, traditionally viewed as nurturing, youth-friendly spaces dedicated to protecting and educating children, have become among the most punitive institutions young people now face—on a daily basis. Educating for citizenship, work, and the public good has been replaced with models of schooling in which students are viewed narrowly—on the one hand as threats or perpetrators of violence, or on the other as infantilized potential victims of crime (on the Internet, at school, and in other youth spheres) who must endure modes of governing that are demeaning and repressive. Jonathan Simon captures this transformation of schools from a public good to a security risk in the following comment:

Today, in the United States, it is crime that dominates the symbolic passageway to school and citizenship. And behind this surface, the pathways of knowledge and power within the school are increasingly being shaped by crime as the model problem, and tools of criminal justice as the dominant technologies. Through the introduction of police, probation officers, prosecutors, and a host of private security professionals into the schools, new forms of expertise now openly compete with pedagogic knowledge and authority for shaping routines and rituals of schools.... At its core, the implicit fallacy dominating many school policy debates today consists of a gross conflation of virtually all the vulnerabilities of children and youth into variations on the theme of crime. This may work to raise the salience of education on the public agenda, but at the cost to students of an education embedded with themes of "accountability," "zero tolerance," and "norm shaping."[86]

The merging of the neoliberal state, in which kids appear as commodities or as a source of profits, and the punishing state, which harkens back to the old days of racial apartheid in its ongoing race to incarcerate, was made quite visible in a recent shocking account of two judges in Pennsylvania who took bribes as part of a scheme to fill up privately run juvenile detention centers with as many youths as possible, regardless of how minor the infraction they committed. One victim, Hillary Transue, appeared before a "kickback" judge for "building a spoof *MySpace* page mocking the assistant principal at her high school."[87] A top student who had never been in trouble, she anticipated a stern lecture from the judge for her impropriety. Instead, he sentenced her "to three months at a juvenile detention center on a charge of harassment."[88] It has been estimated that the two judges, Mark A. Ciavarella Jr. and Michael T. Conahan, "made more than $2.6 million in kickbacks to send teenagers to two privately run youth detention centers" and that over 5000 juveniles have gone to jail since the "scheme started in 2003. Many of them were first time offenders and some remain in detention."[89] While this incident received some mainstream new coverage, most of the response focused less on the suffering endured by the young victims than on the breach of professional ethics by the two judges. None of the coverage treated the incident as either symptomatic of the war being waged against youth marginalized by class and race or an issue that the Obama administration should give top priority in reversing.

As the *New York Times'* op-ed writer, Bob Herbert, points out, "school officials and the criminal justice system are criminalizing children and teenagers all over the country, arresting them and throwing them in jail for behavior that in years past would never have led to the intervention of law enforcement."[90] Young people are being ushered "into the bowels of police precincts and jail cells" for minor offenses, which Herbert argues "is a problem that has gotten out of control ... especially as zero tolerance policies proliferate, children are being treated like criminals."[91] The sociologist Randall Beger has written that the new security culture in school comes with an emphasis on "barbed-wire security fences, banned book bags and pagers ... 'lock down drills' and 'SWAT team' rehearsals."[92] As the logic of the market and "the crime complex"[93] frame a number of social actions in schools, students are subjected to three particularly offensive policies, defended by school authorities and politicians under the rubric of school safety. First, students are increasingly subjected to zero tolerance laws that are used primarily to punish, repress, and exclude them. Second, they are increasingly subjected to a "crime complex" in which security staff using harsh disciplinary practices now displace the normative functions teachers once provided both in and outside of the classroom. Third, more and more schools are breaking down the space between education and juvenile delinquency, substituting penal pedagogies for critical learning and replacing a school culture that fosters a discourse of possibility with a culture of fear and social control. Consequently, many youth of color in urban school systems are not just being suspended or expelled from school but also have to bear the terrible burden of being ushered into the dark precincts of juvenile detention centers, adult courts, and prison.

Once seen as an invaluable public good and a laboratory for critical learning and engaged citizenship, public schools are increasingly viewed as sites of crime, warehouses, or containment centers. Consequently, students are also reconceived through the optic of crime as populations to be managed and controlled primarily by security forces. In accordance with this perception of students as potential criminals and the school as a site of disorder and delinquency, schools across the country since the 1980s have implemented zero tolerance policies that involve the automatic imposition of severe penalties for first offenses of a wide range of undesirable, but often harmless, behaviors.[94] Based on the assumption that schools are rife with crime, and fueled by the emergence of a number of state and federal laws such as the Gun-Free Schools Act of 1994, mandatory sentencing legislation, and the popular "three strikes and you're out" policy, many educators first invoked zero tolerance rules against kids who brought firearms to schools—this was exacerbated by the high-profile school shootings in the mid-1990s. But as the climate of fear increased, the assumption that schools were dealing with a new breed of student— violent, amoral, and apathetic—began to take hold in the public imagination. Moreover, as school safety became a top educational priority, zero tolerance policies were broadened and now include a range of behavioral infractions that encompass everything from possessing drugs or weapons to threatening other students—all broadly conceived. Under zero tolerance policies, forms of punishments that were once applied to adults now apply to first graders. Students who violate what appear to be the most minor rules—such as a dress code violation—are increasingly subjected to zero tolerance laws that have a disparate impact on students of color while being needlessly punitive. The punitive nature of the zero tolerance approach is on display in a number of cases where students have had to face harsh penalties that defy human compassion and reason. For example, an eight-year-old boy in the first grade at a Miami elementary school took a table knife to his school, using it to rob a classmate of $1 in lunch money. School officials claimed he was facing "possible expulsion and charges of armed robbery."[95] In another instance that took place in December 2004, "Porsche, a fourth-grade student at a Philadelphia, PA, elementary school, was yanked out of class, handcuffed, taken to the police station and held for eight hours for bringing a pair of 8-inch scissors to school. She had been using the scissors to work on a school project at home. School district officials acknowledged that the young girl was not using the scissors as a weapon or threatening anyone with them, but scissors qualified as a potential weapon under state law."[96] It gets worse. Adopting a rigidly authoritarian zero tolerance school discipline policy, the following incident in the Chicago public school system signals both bad faith and terrible judgment on the part of educators implementing these practices. According to the report *Education on Lockdown*:

in February 2003, a 7-year-old boy was cuffed, shackled, and forced to lie face down for more than an hour while being restrained by a security officer at Parker Community Academy on the Southwest Side. Neither the principal nor the assistant principal came to the aid of the first grader, who was so traumatized by the event he was not able to return to school.[97]

Traditionally, students who violated school rules and the rights of others were sent to the principal's office, guidance teacher, or another teacher. Corrective discipline in most cases was a matter of judgment and deliberation generally handled within the school by the appropriate administrator or teacher. Under such circumstances, young people could defend themselves, the context of their rule violation was explored (including underlying issues, such as problems at home, that may have triggered the behavior in the first place), and the discipline they received was suited to the nature of the offense. Today, as school districts link up with law enforcement agencies, young people find themselves not only being expelled or suspended at record rates but also being "subject to citations or arrests and referrals to juvenile or criminal courts."[98] Students who break even minor rules, such as pouring a glass of milk on another student or engaging in a schoolyard fight, have been removed from the normal school population, handed over to armed police, arrested, handcuffed, shoved into patrol cars, taken to jail, fingerprinted, and subjected to the harsh dictates of the juvenile and criminal justice systems. As Bernardine Dohrn points out:

Today, behaviors that were once punished or sanctioned by the school vice-principal, family members, a neighbor, or a coach are more likely to lead to an adolescent being arrested, referred to juvenile or criminal court, formally adjudicated, incarcerated in a detention center, waived or transferred to adult criminal court for trial, sentenced under mandatory sentencing guidelines, and incarcerated with adults.[99]

How educators think about children through a discourse that has shifted from hope to punishment is evident in the effects of zero tolerance policies, which criminalize student behavior in ways that take an incalculable toll on their lives and their future. For example, between 2000 and 2004, the Denver public school system experienced a

71 percent increase in the number of student referrals to law enforcement, many for nonviolent behaviors. The Chicago school system in 2003 had over 8000 students arrested, often for trivial infractions such as pushing, tardiness, and using spitballs. As part of a human waste-management system, zero tolerance policies have been responsible for suspending and expelling black students in record-high numbers. For instance, "in 2000, Blacks were 17 percent of public school enrollment nationwide and 34 percent of suspensions."[100] And when poor black youth are not being suspended under the merger of school security and law-and-order policies, they are increasingly at risk of falling into the school-to-prison pipeline. As the Advancement Project points out, the racial disparities in school suspensions, expulsions, and arrests feeds and mirrors similar disparities in the juvenile and criminal justice systems:

> [I]n 2002, Black youths made up 16% of the juvenile population but were 43% of juvenile arrests, while White youths were 78% of the juvenile population but 55% of juvenile arrests. Further, in 1999, minority youths accounted for 34% of the U.S. juvenile population but 62% of the youths in juvenile facilities. Because higher rates of suspensions and expulsions are likely to lead to higher rates of juvenile incarceration, it is not surprising that Black and Latino youths are disproportionately represented among young people held in juvenile prisons.[101]

The city of Chicago, which has a large black student population, implemented a take-no-prisoners approach in its use of zero tolerance policies, and the racially skewed consequences are visible in grim statistics, revealing that "every day, on average, more than 266 suspensions are doled out . . . during the school year." Moreover, the number of expulsions has "mushroomed from 32 in 1995 to 3000 in the school year 2003–2004,"[102] most affecting poor black youth.

As the culture of fear, crime, and repression dominate American public schools, the culture of schooling is reconfigured through the allocation of resources used primarily to acquire more police, security staff, and technologies of control and surveillance. In some cases, schools such as those in the Palm Beach County system have established their own police departments. Saturating schools with police and security personnel has created a host of problems for schools, teachers, and students—not to mention that such policies tap into financial resources otherwise used for actually enhancing learning. In many cases, the police and security guards assigned to schools are not properly trained to deal with students and often use their authority in ways that extend far beyond what is either reasonable or even legal. When Mayor Bloomberg in 1998 allowed control of school safety to be transferred to the New York Police Department, the effect was not only a boom in the number of police and school safety agents but also an intensification of abuse, harassments, and arrests of students throughout the school system.

One example of war-on-terror tactics used domestically and impacting schools can be seen in the use of the roving metal detector program in which the police arrive at a school unannounced and submit all students to metal detector scans. In *Criminalizing the Classroom*, Elora Mukherjee describes some of the disruptions caused by the program:

> As soon as it was implemented, the program began to cause chaos and lost instructional time at targeted schools, each morning transforming an ordinary city school into a massive police encampment with dozens of police vehicles, as many as sixty SSAs [School Security Agents] and NYPD officers, and long lines of students waiting to pass through the detectors to get to class.[103]

As she indicates, the program does more than delay classes and instructional time: it also fosters abuse and violence. The following incident at Wadleigh Secondary School on November 17, 2006, provides an example of how students are abused by some of the police and security guards. Mukherjee writes:

> The officers did not limit their search to weapons and other illegal items. They confiscated cell phones, iPods, food, school supplies, and other personal items. Even students with very good reasons to carry a cell phone were given no exemption. A young girl with a pacemaker told an officer that she needed her cell phone in case of a medical emergency, but the phone was seized nonetheless. When a student wandered out of line, officers screamed, "Get the fuck back in line!" When a school counselor asked the officers to refrain from cursing, one officer retorted, "I can do and say whatever I want," and continued, with her colleagues, to curse.[104]

Many students in New York City have claimed that the police are often disrespectful and verbally abusive, stating that "police curse at them, scream at them, treat them like criminals, and are on 'power trips.'...At Martin Luther King Jr. High School, one student reported, SSAs refer to students as 'baby Rikers,' implying that they are convicts-in-waiting. At Louis D. Brandeis High School, SSAs degrade students with comments like, 'That girl has no ass.' "[105] In some cases, students who had severe health problems had their phones taken away and when they protested were either arrested or assaulted. Mukherjee reports that "[a] school aide at Paul Robeson High School witnessed a Sergeant yell at, push, and then physically assault a child who would not turn over his cell phone. The Sergeant hit the child in the jaw, wrestled him to the ground, handcuffed him, removed him from school premises, and confined him at the local precinct."[106] There have also been cases of teachers and administrators being verbally abused, assaulted, and arrested while trying to protect students from overzealous security personnel or police officers.

Under such circumstances, schools begin to take on the obscene and violent contours one associates with maximum security prisons: unannounced locker searches, armed police patrolling the corridors, mandatory drug testing, and the ever-present phalanx of lock-down security devices such as metal detectors, X-ray machines, surveillance cameras, and other technologies of fear and control. Appreciated less for their capacity to be educated than for the threat they pose to adults, students are now treated as if they were inmates, often humiliated, detained, searched, and in some cases arrested. Randall Beger is right in suggesting that the new "security culture in public schools [has] turned them into 'learning prisons' where the students unwittingly become 'guinea pigs' to test the latest security devices."[107]

Poor black and Latino male youth are particularly at risk in this mix of demonic representation and punitive modes of control as they are the primary object of not only racist stereotypes but also a range of disciplinary policies that criminalize their behavior.[108] Such youth, increasingly viewed as burdensome and dispensable, now bear the brunt of these assaults by being expelled from schools, tried in the criminal justice system as adults, and arrested and jailed at rates that far exceed their white counterparts.[109] While black children make up only 15 percent of the juvenile population in the United States, they account for 46 percent of those put behind bars and 52 percent of those whose cases end up in adult criminal courts. Shockingly, in the land of the free and the home of the brave, "[a] jail or detention cell after a child or youth gets into trouble is the only universally guaranteed child policy in America."[110]

When their behavior is not being criminalized, youth are often held in contempt and treated with cynical disrespect. For example, administrators at Gonzales High School in Texas decided that if students violated the school's highly conservative dress code, they would be treated like convicts and forced to wear prison-style jumpsuits, unless they procured another set of clothes from their parents. Larry Wehde, the superintendent, justified this obvious abuse of school authority by simply restating his own (blind) faith in the reactionary ideology that produced the policy: "We're a conservative community, and we're just trying to make our students more reflective of that."[111] Indeed! With no irony intended, the school board president, Glenn Menking, said the purpose of the code was "to put students' attention on education, not clothes."[112] Neither administrator revealed doubts about establishing school disciplinary practices modeled on prison policies. The critical lesson for students in this instance is to be wary of adults who seem to believe that treating young people like prison inmates is effective training for their entry into the twenty-first century. Not surprisingly, some parents have voiced outrage over the policy, stating that their children should not be treated like "little prisoners."[113]

That students are being miseducated, criminalized, and arrested through a form of penal pedagogy in lock-down schools that resemble prisons is a cruel reminder of the degree to which mainstream politicians and the American public have turned their backs on young people in general and poor minority youth in particular. As schools are reconfigured around the model of the prison, crime becomes the central metaphor used to define the nature of schooling, while criminalizing the behavior of young people becomes the most valued strategy in mediating the relationship between educators and students. The consequences of these policies for young people suggest not only an egregious abdication of responsibility—as well as reason, judgment, and restraint—on the part of administrators, teachers, and parents but also a new role for schools as they become more prisonlike, eagerly adapting to their role as an adjunct of the punishing state.

As schools define themselves through the lens of crime and merge with the dictates of the penal system, they eliminate a critical and nurturing space in which to educate and protect children in accordance with the ideals of a democratic society. As central institutions in the youth disposability industry, public schools now serve to discipline and warehouse youth, while they also put in place a circuit of policies and practices to make it easier for minority youth to move from schools into the juvenile justice system and eventually into prison. The combination of school punishments and criminal penalties has proven a lethal mix for many poor minority youth and has transformed schools from spaces of youth advocacy, protection, hope, and equity to military fortresses, increasingly well positioned to mete out injustice and humiliation, transforming the once-nurturing landscapes that young people are compelled to inhabit. Rather than confront the war on youth, especially the increasing criminalization of their behavior, schools now adopt policies that both participate in and legitimate the increasing absorption of young people into the juvenile and adult criminal justice system. Commenting on the role of schools as a major feeder of children into the adult criminal court system, Bernardine Dohrn writes:

> As youth service systems (schools, foster care, probation, mental health) are scaling back, shutting down, or transforming their purpose, one system has been expanding its outreach to youth at an accelerated rate: the adult criminal justice system. All across the nation, states have been expanding the jurisdiction of adult criminal court to include younger children by lowering the minimum age of criminal jurisdiction and expanding the types of offenses and mechanisms for transfer or waiver of juveniles into adult criminal court. Barriers between adult criminals and children are being removed in police stations, courthouses, holding cells, and correctional institutions. Simultaneously, juvenile jurisdiction has expanded to include both younger children and delinquency sentencing beyond the age of childhood, giving law enforcement multiple options for convicting and incarcerating youngsters.[114]

Although state repression aimed at children is not new, what is unique about the current historical moment is that the forces of domestic militarization are expanding, making it easier to put young people in jail rather than to provide them with the education, services, and care they need to face the growing problems characteristic of a democracy under siege. As minority youth increasingly become the objects of severe disciplinary practices in public schools, many often find themselves vulnerable and powerless as they are thrown into juvenile and adult courts, or even worse, into overcrowded and dangerous juvenile correctional institutions and sometimes adult prisons.[115]

There is a special level of danger and risk that young people face when they enter the criminal justice system in the United States, and the figures are staggering. For example, one recent report states, "These systems affect a wide swath of the U.S. youth population. Nationwide each year, police make 2.2 million juvenile arrests; 1.7 million cases are referred to juvenile courts; an estimated 400,000 youngsters cycle through juvenile detention centers; and nearly 100,000 youth are confined in juvenile jails, prisons, boot camps, and other residential facilities on any given night."[116] The tragedy is that some of these youth are sentenced to die in prisons. For instance, a report issued by the Equal Justice Initiative in 2007 states, "In the United States, dozens of 13- and 14-year-old children have been sentenced to life imprisonment with no possibility of parole after being prosecuted as adults."[117] In this case, the United States has the dubious distinction of being the only country in the world "where a

13-year-old is known to be sentenced to life in prison without the possibility of parole."[118] What is to be said about a country that is willing to put young children behind bars until they die? These so-called criminals are not adults; they are immature and underdeveloped children who are too young to marry, drive a car, get a tattoo, or go to scary movies, but allegedly not too young to be put in prison for the rest of their lives. According to a recent Equal Justice Initiative report, "at least 2225 people are serving sentences of death in prison for crimes they committed under the age of 18," including "73 children who are either 13- or 14-years-old."[119] Moreover, on any given day in the United States, "9500 juveniles under the age of 18 are locked up in adult penal institutions."[120] At the current time, 44 states and the District of Columbia can try 14-year-olds in the adult criminal system.[121]

Giving up on the idea of rehabilitation is bad enough when applied to incarcerated adults, but it is unforgivable when applied to children. Not only do young people who find themselves in adult prisons have few opportunities for acquiring meaningful work skills and getting a decent education, they are also at great risk for physical and sexual assault. As the Equal Justice Initiative report points out:

Juveniles placed in adult prisons are at heightened risk of physical and sexual assault by older, more mature prisoners. Many adolescents suffer horrific abuse for years when sentenced to die in prison. Young inmates are at particular risk of rape in prison. Children sentenced to adult prisons typically are victimized because they have "no prison experience, friends, companions or social support." Children are five times more likely to be sexually assaulted in adult prisons than in juvenile facilities.[122]

And when they are removed from the adult prison population, youth are often placed in isolation, locked down "23 hours a day in small cells with no natural light."[123] One consequence of placing young people in these environments is that these punitive conditions "exacerbate existing mental disorders, and increase risk of suicide. In fact, youth have the highest suicide rates of all inmates in jails. Youth are 19 times more likely to commit suicide in jail than youth in the general population and 36 times more likely to commit suicide in an adult jail than in a juvenile detention facility. Jail staff are simply not equipped to protect youth from the dangers of adult jails."[124]

Such cruel and unusual punishment is borne disproportionately by poor minority youth. In fact, "Of the 73 children between the ages of 13–14 years-old sentenced to die in prison, nearly half (36, or 49%) are African American. Seven (9.6%) are Latino. Twenty-two

(30%) are white.... [while] all of the children condemned to death in prison for non-homicide offenses are children of color. All but one of the children sentenced to life without the possibility of parole for offenses committed at age 13 are children of color."[125] Unfortunately, the children who increasingly inhabit juvenile courts, adult courts, and correctional facilities in the United States emerge from a public school system that has been severely undermined as a democratic public sphere. Subject to harsh market forces, cutbacks in already meager state budgets, the disdain of neoconservative policies, and the massive disempowerment of teachers by an audit-and-testing culture, the public schools in the United States have defaulted on their responsibility to young people. What is at stake in governance under the punishing state is made clear, once again, by Bernardine Dohrn. She writes:

Criminalizing youth behaviors, policing schools, punishing children by depriving them of an education, constricting social protections for abused and neglected youth, and subjecting youth to law enforcement as a "social service"—these trends smack of social injustice, racial inequity, dehumanization, and fear-filled demonization of youngsters, who are our prospective hope. At stake here is the civic will to invest in our common future by seeing other people's children as our own.

Clearly, any attempt to invest in a common and just future implies that educators bear some of the responsibility for the terrible injustices and extraordinary abuse minority youth are experiencing in the United States under a political and economic mode of governance that holds them in contempt while it simultaneously makes them disposable. Educators and others can work to reverse the kinds of policies and practices that emerge from the current war on kids by making visible the interlocking ideologies and practices in which incarceration and punishment become a substitute for "early intervention and sustained child investment."[126] Similarly, policies will have to be put into place that not only remove young people from jails but also vastly reduce the number of young people who enter the child welfare and juvenile and criminal justice systems. At the very least, such a task suggests reforming those primary institutions such as schools, the mainstream media, and the criminal justice system that not only demonize and punish youth but also play a pivotal role in pushing them into the disciplinary apparatuses of the punishing state, especially mass incarceration. Any viable politics aimed at improving the lives of young people will also have to address what it means to challenge those commanding institutions whose priorities for the last 30 years suggest to poor and minority youth they are not worthy of the best future that the richest democracy in the world has to offer them. What must be challenged and reversed is the all-too-common assumption that American society is more willing to invest in sending them to jail than in providing them with high-quality schools, decent education, and the promise of a better life.

Such a task is formidable, and there is more at stake here than creating a society that provides a level playing field for all children and youth, a society in which matters of equality and justice trump the needs of markets and a rationality of excessive self-interest. As Lawrence Grossberg has argued, there is also the need for educators and others "to reimagine imagination itself—not only visions of an alternative future, but also new languages of possibility and new understandings of an act of envisioning a better future."[127] It is difficult to imagine what it means to fight for the rights of children, if we cannot at the same time imagine a different conception of the future, one vastly at odds with a present that can only portend a future as a repeat of itself. But living in the shadow of a vicious realignment of a punishing state and a ruthless mode of economic Darwinism also demands more than a commitment to justice, democratic values, and hope: it necessitates the hard work of building social movements willing to push dominant relations of power over the tipping point in order to make good for children the promise of a real democracy. Within this current moment of uncertainty and possibility, it is necessary for educators, artists, intellectuals, and others to raise questions and develop rigorous modes of analysis in order to explain how a culture of domestic militarization, with its policies of containment and brutalization, has been able to develop and gain consent from so many people in the United States during the last three decades. And, most importantly, such a challenge suggests rethinking the possibility of a new mode of politics and empowering forms of education, especially in light of the Obama victory, that work and struggle vigorously for a social order willing to expand and strengthen the ideals and social relations of a more just society, one in which a future of hope and imagination is inextricably connected to the fate of all young people, if not democracy itself. Although the Obama administration has pledged billions to early childhood education, Obama's appointment of Arne Duncan to the education cabinet position is a deep cause of worry for many educators. Given Duncan's track record in Chicago, where he was a staunch advocate for harsh zero tolerance policies, endorsed a now-discredited business model for schools, and supported data-driven instruction, merit pay, standardized testing, charter schools, and most disturbingly paying students to consume digestible knowledge, educators and others can waste no time organizing social movements willing to struggle for democratic reforms that enable critical learning, produce access to quality schools for all students, and deepen democratic values, rather than close them down.[128]

Under this insufferable climate of increased repression and unabated exploitation, young people and communities of color become the new casualties in an ongoing war against justice, freedom, social citizenship, and democracy. Given the switch in public policy from social investment to punishment—a policy that in education, for now, Obama seems willing to support—it is clear that young people for whom race and class loom large have become disposable. How much longer can a nation ignore those youth who lack the resources and opportunities that were available, in a partial and incomplete way, to previous generations? And what does it mean when a nation becomes frozen ethically and imaginatively in providing its youth with a future of hope and opportunity? Under such circumstances, it is time for intellectuals who inhabit a wider variety of public spheres to take a stand and to remind themselves that collective problems deserve collective solutions and that what is at risk is not only a generation of young people and adults now considered to be a generation of suspects, but the very possibility of deepening and expanding democracy itself.

# CHAPTER 3

# LOCKED OUT: YOUTH AND ACADEMIC UNFREEDOM

*Democracy is not an institution, but essentially an anti-institutional force, a "rupture" in the otherwise relentless trend of the powers-that-be to arrest change, to silence and to eliminate from the political process all those who have not been "born" into power. . . . Democracy expresses itself in a continuous and relentless critique of institutions; democracy is an anarchic, disruptive element inside the political system; essentially, a force for* dissent *and change. One can best recognize a democratic society by its constant complaints that it is* not *democratic enough.*

—Zygmunt Bauman, *The Individualized Society*[1]

## HIGHER EDUCATION UNDER SIEGE

As corporate power, right-wing think tanks, and military interests jointly engage in an effort to take over higher education, the resistance of educational and other democratic public spheres to a growing anti-intellectualism in American life seems to be weakening. Youth and critical education are the first casualties in the war being waged to force universities and colleges to abandon their autonomy along with their critical role in questioning and promoting the conditions that foster democracy. Instead of serving students and young people, who collectively represent the purpose and future of both education and democracy in the United States, higher education is increasingly administered in a corporate fashion, not only enabling a growing elitism by raising tuition fees but also dangerously embracing a narrow set of interests that put at risk the future of young people, education, and the nation as a whole. Scholarships and programs that enable disadvantaged students to attend and graduate from university and college have been ruthlessly cut back or tied to military service. As higher education increasingly becomes a privilege rather than a right, many working-class youth either find it financially impossible to enter college or, because of increased costs, drop out.[2] Those students who have the resources to stay in school are feeling the pressure of the job market, increasingly so under the current recession, and rush to take courses and receive professional credentials in business and the biosciences as the humanities lose majors and downsize.[3] Under the strain of the current financial crisis, "the rising cost of college threatens to put higher education out of reach for most Americans."[4] While the education gap in the United States has been widening for some time, it is being exacerbated by a wealth gap directly tied to the structural inequities fundamental to a social order shaped by the market-driven politics of neoliberalism. While the middle class will be greatly affected by such costs, it is poor and working-class kids who will find they have almost no chance to attend college, further solidifying their status as redundant and expendable. At a time when youth are increasingly constructed and treated as a disposable population, the university needs to play a role in fighting for the future of all young people rather than a privileged few and for the democratic principles and opportunities that will enable them to be active, critical citizens.

Central to higher education's defense of public responsibility and participation in democratic self-governance is revitalizing its commitment to academic freedom. At one time in history, it may have been unthinkable that university classrooms would be subject to ideological oversight, largely promoted through the interests of outside conservative politicians, foundations, and media. But as more and more teaching positions are contracted out to part-time faculty who have no governance role in the university, and university administrators increasingly succumb to external pressures and intimidation tactics used by conservative think tanks, which actively engage in scanning university departments and classrooms for what they consider left or liberal viewpoints, the classroom is no longer a safe space immune from the corporate and ideological battles being waged and lost at institutional and social levels to a host of neoliberal and right-wing forces.

Overworked and subject to corporate-minded policies imposed by university administrators, many educators are turning away from their responsibility as critically engaged intellectuals, hoping to remain secure in their jobs by blending into the background, minimalizing their personal and political investments by viewing themselves as detached professionals, and reducing classroom teaching to a mere vocational exercise. Yet several recent cases of universities denying tenure to or firing accomplished scholars and teachers for what is seen as their dissident political views are indicative of an ominous future in which academic positions afford little or no security and the content of research and teaching are tightly controlled and censored by institutional mercenaries who reduce education to a business to be managed in the most cost-effective, consumer-oriented terms. Moreover, some colleges are using the current financial downturn to argue for the elimination of tenure, allegedly as a cost-saving measure, thus promoting their conservative ideology and dislike for shared governance under the pretext of a neoliberal call to efficiency.[5] Defending the autonomy of teachers and promoting critical forms of education have become inextricable from defending higher education and the rights of young people to quality education. This chapter explores issues confronting higher education with the purpose of reaffirming its significance as both a foundation for society's investment in young people and the sustenance of democracy itself.

Academics, at the very least, have a moral and political obligation to stand up against the anti-democratic forces attacking higher education, to acknowledge that educational institutions wield enormous cultural power and influence, and to identify with their ethical obligation to assert their cultural authority in ways that foster open-mindedness, dialogue, critical thinking, political agency, and public responsibility. Education is the heart of the democratic political life, and the students and professors who people the campuses of universities and colleges are the heart of higher education. What does it mean when a 2008 study entitled *Closed Minds? Politics and Ideology in American Universities* found that "universities generally have all but ignored what used to be called civics and civic education"?[6] As higher education risks abdicating its role as a democratic public sphere, the hope for a better future for today's youth and the means to fight against the biopolitics of disposability are lost. It is the responsibility of educators, students, parents, labor, and various social movements to organize a collective challenge against higher education's irresponsible and morally indefensible wagering of both young people's futures and the democratic foundations of governance. If left unchecked, the university will be transformed in short order by policies that objectify students and teachers as mere place fillers and reduce learning to a commodity whose value is measured in terms of how it provides economic success rather than how it models the skills to think critically and participate in democratic processes. Nothing less than the lives of young people and the future of democracy is at stake.

Educating young people in the spirit of a critical democracy by providing them with the knowledge, passion, civic capacities, and social responsibility necessary to address the problems facing the nation and globe has always been challenged by the existence of rigid disciplinary boundaries, the cult of expertise or highly specialized scholarship unrelated to public life, and antidemocratic ideologies that scoff at the exercise of academic freedom.[7] Such antidemocratic and anti-intellectual tendencies have intensified in recent decades alongside the contemporary emergence of a number of diverse fundamentalisms, including a market-based neoliberal rationality that exhibits a deep disdain, if not outright contempt, for both democracy and publicly engaged teaching and scholarship. In such circumstances, it is not surprising that academia in the United States is often held hostage to political and economic forces that wish to convert educational institutions into corporate establishments defined by a profit-oriented identity and mission. This means that while the American university still employs the rhetoric of a democratic public sphere, there is a growing gap between a stated belief in noble purposes and the reality of an academy that is under siege.

In keeping with the progressive impoverishment of politics and public life over the past three decades, the university is being transformed into a training ground for corporate and military employment and a cheerleader for a reactionary notion of patriotic correctness, rather than being a public sphere in which youth can become the critical citizens and democratic agents necessary to nourish a socially responsible future. Strapped for money and increasingly defined in the language of a militarized and corporate culture, many universities are now part of an unholy alliance that largely serves the interests of the national security state and the policies of transnational corporations while increasingly removing academic knowledge production from democratic values and projects.[8] College presidents are now called CEOs and speak largely in the discourse of Wall Street and corporate fund managers. Venture capitalists scour colleges and universities in search of big profits to be made through licensing agreements, the control of intellectual property rights, and investments in university spin-off companies. In this new, though recently humbled, Gilded Age of money and profit, academic subjects gain stature almost exclusively through their exchange value on the market. It is also true that students who have scrambled to get MBAs are now taking government and public service jobs as employment opportunities in the banking and financial sectors are drying up.

Not surprisingly, students are now referred to as "customers," while some university presidents even argue that professors should be labeled "academic entrepreneurs."[9] Tenured faculty are called upon to generate grants, establish close partnerships with corporations, and teach courses that have practical value in the marketplace. There is little in this vision of the university that imagines young people as anything other than fodder for the corporation or appendages of the national security state. What was once the hidden curriculum of many universities—the subordination of higher education to capital—has now become an open and much-celebrated policy of both public and private higher education. As higher education is corporatized, young people find themselves on campuses that look more like malls and they are increasingly taught by professors who are hired on a contractual basis, have obscene workloads, and barely make enough money to pay off their student loans. Worth noting is that "both part-time and full-timers not on a tenure track account for nearly 70 percent of professors at colleges and universities, both public and private."[10]

Higher education is increasingly abandoning its faith in and commitment to democracy as it aligns itself with corporate power and military values, while at the same time succumbing to a range of right-wing religious and political attacks.[11] Instead of being a space of critical dialogue, analysis, and interpretation, it is increasingly defined as a space of consumption, further marginalizing young people without access to financial resources and validating ideas in instrumental terms, linked for example to the ability to attract outside funding. As the university develops increasingly "strong ties with corporate and warfare powers,"[12] the culture of research is oriented toward the needs of the military-industrial-academic complex. Faculty and students find their work further removed from the language of democratic values and their respective roles modeled largely upon the business entrepreneur, the consumer, or the soldier in the "war on terror." With no irony intended, Professor Philip Leopold argues that it is an "essential part of an academic career" that academics be viewed as business entrepreneurs, trained to "watch the bottom line" and to be attentive to "principles of finance, management, and marketing" and to the development of a "brand identity (academic reputation) that is built on marketing (publications and presentations) of a high-quality product (new knowledge)."[13] In another statement pregnant with irony, Robert Gates, the secretary of defense, proposed the creation of what he calls a new "Minerva consortium," ironically named after the goddess of wisdom, whose purpose is to fund various universities to "carry out social-sciences research relevant to national security."[14] Gates and others would like to turn universities into militarized knowledge factories more willing to produce knowledge, research, and personnel in the interest of the warfare and Homeland (In)Security State than to assume the important role of tackling the problems of contemporary life while holding dominant institutions—especially those that trade in force, violence, and militarism—accountable by questioning how their core values and presence in the world alter and shape democratic identities, values, and organizations. Since September 11, 2001, the CIA and other agencies have been a growing presence on American campuses, offering federal scholarship programs, grants, and other forms of financial aid to students in exchange for postgraduate service within the intelligence or military agencies.[15] Such incursions by governmental and corporate interests have become highly influential in shaping the purpose and meaning of higher education. Unfortunately, Gates' view of the university as a militarized knowledge factory and Professor Leopold's instrumental understanding of the university as a new marketplace of commerce now parade under the banner of educational reform and produce little resistance from either the public or academics. Even the allegedly liberal Obama administration has bought into this morally disdainful understanding of the meaning and purpose of higher education.[16] Hence, it no longer seems unreasonable to argue that just as democracy is being emptied out, the university is also being stripped of its role as a democratic setting where, though often in historically fraught ways, a democratic ethos has been cultivated, practiced, and sustained for several generations.

Higher education is increasingly being influenced by larger economic, military, and ideological forces that consistently attempt to narrow its legitimacy and purview as a democratic public sphere. Public intellectuals are now replaced by privatized intellectuals often working in secrecy and engaged in research that serves either the warfare state or the corporate state, or both. Intellectuals are no longer placed in a vibrant relationship to public life but now labor under the influence of managerial modes of governance and market values that mimic the logic of Wall Street. As Jennifer Washburn observes,

In the classroom deans and provosts are concerned less with the quality of instruction than with how much money their professors bring in. As universities become commercial entities, the space to perform research that is critical of industry or challenges conventional market ideology—research on environmental pollution, poverty alleviation, occupational health hazards—has gradually diminished, as has the willingness of universities to defend professors whose findings conflict with the interests of their corporate sponsors. Will universities stand up for academic freedom in these situations, or will they bow to commercial pressure out of fear of alienating their donors?[17]

As a consequence of this pressure, higher education appears to be increasingly decoupling itself from its historical legacy as a crucial public sphere, responsible for both educating students for the workplace and providing them with the modes of critical discourse, interpretation, judgment, imagination, and experiences that deepen and expand democracy. As universities adopt the ideology of the transnational corporation and become subordinated to the needs of capital, the war industries, and the Pentagon, they are less concerned about how they might educate students about the ideology and civic practices of democratic governance and the necessity of using knowledge to address the challenges of public life.[18] Instead, as part of the post-9/11 military-industrial-academic complex, higher education increasingly conjoins military interests and market values, identities, and social relations while John Dewey's once-vaunted claim that "democracy needs to be reborn in each generation, and education is its midwife" is either willfully ignored, forgotten, or made an object of scorn.[19]

The corporatization, militarization, and dumbing down of rigorous scholarship and the devaluing of the critical capacities of young people mark a sharp break from a once-strong educational tradition in the United States, extending from Thomas Jefferson to John Dewey to W. E. B. DuBois, that held that freedom flourishes in the worldly space of the public realm only through the work of educated, critical citizens. Within this democratic tradition, education was not confused with training, nor did it surrender its democratic values to an unquestioning faith in market efficiency; instead, its critical function was propelled by the need to provide students with the knowledge and skills that enable a "politically interested and mobilized citizenry, one that has certain solidarities, is capable of acting on its own behalf, and anticipates a future of ever greater social equality across lines of race, gender, and class."[20] Other prominent educators and theorists such as Hannah Arendt, James B. Conant, and Cornelius Castoriadis have long believed and rightly argued that we should not allow education to be modeled after the business world. Dewey, in particular, warned about the growing influence of the "corporate mentality" and the threat that the business model posed to public spaces, higher education, and democracy. He argued:

The business mind [has] its own conversation and language, its own interests, its own intimate groupings in which men of this mind, in their collective capacity, determine the tone of society at large as well as the government of industrial society.... We now have, although without formal or legal status, a mental and moral corporateness for which history affords no parallel.[21]

Dewey and the other public intellectuals mentioned above shared a common vision and project of rethinking what role education might play in providing students with the habits of mind and ways of acting that would enable them to "identify and probe the most serious threats and dangers that democracy faces in a global world dominated by instrumental and technological thinking."[22] James Bryant Conant, a former president of Harvard University, argued that higher education should create a class of "American radicals" who could fight for equality, favor public education, elevate human needs over property rights, and challenge "groups which have attained too much power."[23] Conant's views seem so radical today that it is hard to imagine him being hired as a university president at Harvard or at any other institution of higher learning. All of these intellectuals offered a notion of the university as a bastion of democratic learning and values that provides a crucial referent in exploring the more specific question regarding what form the relationship between corporations and higher education will take in the twenty-first century. It now seems naïve to assume that corporations, left to their own devices, would view higher education as more than merely a training center for future business employees, a franchise for generating profits, or a space in which corporate culture and education merge in order to produce literate consumers.

American higher education is ever more divided into those institutions that educate the elite to rule the world in the twenty-first century and the second- and third-tier institutions that largely train students for low-paid positions in the capitalist world economy. It is increasingly apparent that the university in America has become a social institution that not only fails to address inequality in society but also contributes to a growing division between social classes. Instead of being a space of critical dialogue, analysis, and interpretation, the American university is increasingly defined as a space of consumption, where ideas are validated in instrumental terms and valued for their success in attracting outside funding while developing stronger ties to corporate powers. Those transcendent values necessary to sustain a democratic society and "nurture the capacity for individual conscience" and critical agency are increasingly being subordinated to a corporatism that crushes "the capacity for moral choice."[24] Moreover, as tuition exceeds the budgets of most Americans, quality education at public and private universities becomes a reserve primarily for the children of the rich and powerful. While researchers attempt to reform a "broken" federal student financial aid system, there is "growing evidence . . . that the United States is slipping (to 10th now among industrialized countries) in the proportion of young adults who attain some postsecondary education."[25]

Higher education has a responsibility not only to be available and accessible to all youth but also to educate young people to make authority politically and morally accountable and to expand both academic freedom and the possibility and promise of the university as a bastion of democratic inquiry, values, and politics, even as these are necessarily refashioned at the beginning of the new millennium. While questions regarding whether the university should serve public rather than private interests no longer carry the weight of forceful criticism that they did when raised by Thorstein Veblen, Robert Lynd, and C. Wright Mills in the first part of the twentieth century, such questions are still crucial in addressing the reality of higher education and what it might mean to imagine the university's full participation in public life as the protector and promoter of democratic values among the next generation. This is especially true at a time when the meaning and purpose of higher education are under attack by a phalanx of right-wing forces attempting to slander, even vilify, liberal and left-oriented professors, cut already meager federal funding for higher education, and place control of what is taught and said in classrooms under legislative oversight.[26] While the American university faces a growing number of problems that range from the increasing loss of federal and state funding to the incursion of corporate power, a galloping commercialization, and the growing influence of the national security state, it is also currently being targeted by conservative forces that have highjacked political power and waged a focused campaign against the principles of academic freedom, sacrificing the quality of education made available to youth in the name of patriotic correctness and dismantling the university as a site of critical pedagogical practice, autonomous scholarship, independent thought, and uncorrupted inquiry.

## THE RIGHT-WING ASSAULT ON HIGHER EDUCATION

Conservatives have a long history of viewing higher education as a cradle of left-wing thought and radicalism. Moreover, just as religious fundamentalists attempted to suppress academic freedom in the nineteenth century, they continue to do so today. Yet in its current expression, the attack on the university has taken a strange turn: liberal professors, specifically in the arts, humanities, and social sciences, are now being portrayed as the enemies of academic freedom because they allegedly abuse students' rights by teaching views unpopular to some of the more conservative students. The current attack on academe borrows its tactics from right-wing strategists who emphasize the power and political nature of education. This viewpoint has been significant in shaping long-term strategies put into place as early as the 1920s to win an ideological war against liberal intellectuals, who instead argued both for changes in American domestic and foreign policy and for holding government and corporate power accountable as a precondition for extending and expanding the promise of an inclusive democracy. During the McCarthy era, criticisms of the university and its dissenting intellectuals cast a dark cloud over the exercise of academic freedom, and many academics were either fired or harassed out of their jobs because of their political activities outside the classroom or their alleged communist fervor or left-wing affiliations. In 1953, the Intercollegiate Studies Institute (ISI) was founded by Frank Chodorov in order to assert right-wing influence and control over universities. ISI was but a precursor to the present era of politicized and paranoid academic assaults. In fact, William F. Buckley, who catapulted to fame among conservatives in the early 1950s with the publication of *God and Man at Yale*, in which he railed against secularism at Yale University and called for the firing of socialist professors, was named as the first president of ISI. The current president of ISI, T. Kenneth Cribb, Jr., delivered the following speech to the Heritage Foundation in 1989 that captures the ideological spirit and project behind its view of higher education:

We must . . . provide resources and guidance to an elite which can take up anew the task of enculturation. Through its journals, lectures, seminars, books and fellowships, this is what ISI has done successfully for 36 years. The coming of age of such elites has provided the current leadership of the conservative revival. But we should add a major new component to our strategy: the conservative movement is now mature enough to sustain a counteroffensive on that last Leftist redoubt, the college campus. . . . We are now strong enough to establish a contemporary presence for conservatism on campus, and contest the Left on its own turf. We plan to do this greatly by expanding the ISI field effort, its network of campus-based programming.[27]

ISI was an early effort on the part of conservatives to " 'take back' the universities from scholars and academic programs regarded either as too hostile to free markets or too critical of the values and history of Western civilization."[28] As part of an effort to influence future generations to adopt a conservative ideology and leadership roles in "battling the radicals and PC types on campus," the Institute now provides numerous scholarships, summer programs, and fellowships to students.[29] The *Chronicle of Higher Education* reported in 2007 that various conservative groups are spending over $40 million "on their college programs."[30] Tying ideology to student funding is dangerous, if not unethical. It enables right-wing organizations to take advantage of low-income families in an attempt to rear up a new generation of conservatives. More recently, conservative foundations are trying to establish "academic beachheads" for their ideas by funding programs, centers, and institutes, largely run by conservative professors. The journalist Patricia Cohen has written that decades of money from conservative foundations have "helped create a kind of shadow university of private research institutes."[31]

Perhaps the most succinct statement for establishing a theoretical framework and political blueprint for the current paranoia surrounding the academy is the Powell Memo, released on August 23, 1971, and authored by Lewis F. Powell, Jr., who would later be appointed as a member of the U.S. Supreme Court. Powell identified the American college campus "as the single most dynamic source" for producing and housing intellectuals "who are unsympathetic to the [free] enterprise system."[32] He recognized that one crucial strategy in changing the political composition of higher education was to convince university administrators and boards of trustees that the most fundamental problem facing universities was the lack of conservative educators, or what he labeled the "imbalance of many faculties."[33] The Powell Memo was designed to develop a broad-based strategy not only to counter dissent but also to develop a material and ideological infrastructure with the capability to transform the American public consciousness through a conservative pedagogical commitment to reproduce the knowledge, values, ideology, and social relations of the corporate state. The Powell Memo, while not the only influence, played an important role in generating, in the words of Lewis Lapham, a "cadre of ultraconservative and self-mythologising millionaires bent on rescuing the country from the hideous grasp of Satanic liberalism."[34] The most powerful members of this group were Joseph Coors in Denver, Richard Mellon Scaife in Pittsburgh, John Olin in New York City, David and Charles Koch in Wichita, the Smith Richardson family in North Carolina, and Harry Bradley in Milwaukee—all of whom agreed to finance a number of right-wing foundations to the tune of roughly $3 billion[35] over 30 years, building and strategically linking "almost 500 think tanks, centers, institutes and concerned citizens groups both within and outside of the academy. . . . A small sampling of these entities includes the Cato Institute, the Heritage Foundation, the American

Enterprise Institute, the Manhattan Institute, the Hoover Institution, the Claremont Institute, the American Council of Trustees and Alumni, Middle East Forum, Accuracy in Media, and the National Association of Scholars."[36] For several decades, right-wing extremists have labored to put into place an ultraconservative reeducation machine—an apparatus for producing and disseminating a public pedagogy in which everything tainted with the stamp of liberal origin and the word "public" would be contested and destroyed.

Given the influence and resources of this long campaign against progressive institutions and critical thought in the United States, it is all the more important that current educators of the next generation of citizens sit up and take notice, especially since the university is one of the few places left where critical dialogue, debate, and dissent can take place. Some theorists believe that not only has the militarization and neoliberal reconstruction of higher education proceeded steadily within the last 25 years, but it is now moving at an accelerated pace, subjecting the academy to what many progressives argue is a new and more dangerous threat. One of the most noted historians of the McCarthy era, Ellen Schrecker, insists that "today's assault on the academy is more serious" because "[u]nlike that of the McCarthy era, it reaches directly into the classroom."[37] As Schrecker suggests, the new war being waged against higher education is not simply against dissenting public intellectuals and academic freedom: it is also deeply implicated in questions of power across the university, specifically regarding who controls the hiring process, the organization of curricula, and the nature of pedagogy itself. The expanding influence of conservative trustees and academics within the university is facilitated by the assistance they receive from a growing number of well-funded and powerful right-wing agencies and groups outside the walls of the academy. Joel Beinin argues that many of these right-wing foundations and institutions have to be understood both as part of a political movement that shapes public knowledge in ways unconstrained by the professional standards of the university and as part of a backlash against the protest movements of the 1960s—which called into question the university as a "knowledge factory" and criticized its failure to take its social functions seriously. He writes:

The substantial role of students and faculty members in the anti–Vietnam War movement; the defection of most university-based Latin America specialists from U.S. policy in the Reagan years, if not earlier; similar, if less widespread, defections among Africa and Middle East specialists; and the "culture wars" of the 1980s and 1990s all contributed to the rise of think tanks funded by right-wing and corporate sources designed to constitute alternative sources of knowledge unconstrained by the standards of peer review, tolerance for dissent, and academic freedom.[38]

Subject to both market mechanisms and right-wing ideological rhetoric about using the academy to defend the values of Western civilization and reinforce the dominant social order, the opportunity to assert the university as a space where young people can be exposed to and explore challenging new ideas appears to be dwindling.

While it is crucial to recognize that the rise of a "new McCarthyism" cannot be attributed exclusively to the radical curtailment of civil liberties initiated by the George W. Bush administration after the cataclysmic events of September 11, 2001, it is nonetheless true that a growing culture of fear and jingoistic patriotism emboldened a post-9/11 patriotic correctness movement, most clearly exemplified by actions of the right-wing American Council of Trustees and Alumni (ACTA), which issued a report shortly after the attacks accusing a supposedly unpatriotic academy of being the "weak link in America's response to the attack."[39] Individuals and groups who opposed George W. Bush's foreign and domestic policies were put on the defensive—some overtly harassed—as right-wing pundits, groups, and foundations repeatedly labeled them "traitors" and "un-American." In some cases, conservative accusations that seemed disturbing, if not disturbed, before the events of 9/11 now appeared perfectly acceptable, especially in the dominant media. The legacy of this new-style McCarthyism was also on display in Ohio, California, and a number of other states where some public universities were requiring job applicants to sign statements confirming that they do not belong to any terrorist organization, as defined by the Bush-Cheney administration, which would basically encompass any organization that voiced opposition to the administration's domestic and foreign policies.

In the aftermath of 9/11, universities were castigated as hotbeds of left-wing radicalism, while conservative students alleged that they were being humiliated and discriminated against in college and university classrooms all across the country. The language and tactics of warfare moved easily between so-called rogue states such as Iraq and a critique of universities whose defense of academic freedom did not sit well with academic and political advocates of the neoliberal security-surveillance state.[40] McCarthy-like blacklists were posted on the Internet by right-wing groups such as Campus Watch, ACTA, and Target of Opportunity,[41] attempting to both out and politically shame allegedly radical professors who were giving "aid and comfort to the enemy" because of their refusal to provide unqualified support for the Bush administration. The nature of conservative acrimony may have been marked by a new language, but the goal of the attack on higher education was largely the same: to remove from the university all vestiges of dissent and to reconstruct it as an increasingly privatized sphere for reproducing the interests of corporations and the national security state while also having it assume a front-line position in the promotion of an imperialist military agenda. "Academic balance" was now invoked as a way to protect American values and national identity when it really promoted a form of affirmative action for hiring conservative faculty. In a similar manner, "academic freedom" was redefined, both through the prism of student rights and as a legitimating referent for dismantling professional academic standards and imposing outside political oversight on the classroom. If the strategy and project of conservative ideologues became more energetic and persistent after 9/11, it is also fair to say that right-wing efforts and demands to reform higher education took a dangerous turn that far exceeded the threat posed by the previous culture wars.

Under the Bush-Cheney administration, the war on terror became a pretext for a war against any public sphere that took responsibility for the welfare of its citizens and residents, including higher education. The neoliberal mantra of "privatize or perish" became a battle cry for a generation of right-wing activists attempting to dismantle public and higher education as democratic public spheres. The right-wing coalition of Christian evangelicals, militant nationalists, market fundamentalists, and neoconservatives that had gained influence under the Reagan administration had unprecedented power in shaping policy under the second Bush presidency. Many academics as well as public school teachers who critically addressed issues such as the U.S. presence in Iraq, the neoconservative view of an imperial presidency, the unchecked market fundamentalism of the Bush administration, or the right-wing views driving energy policies, sex education, and the use of university research "in pursuit of enhanced war making abilities"[42] were either admonished, labeled un-American, or simply fired. Some of the most famous cases include professors such as Joseph Massad of Columbia University, Norman Finkelstein of DePaul University, Nadia Abu E-Haj of Barnard College, and Ward Churchill of the University of Colorado. Though these cases received wide attention in the dominant media, they represent just some of the better-known instances in which academics have been attacked by

right-wing interests through highly organized campaigns of intimidation, which taken collectively suggest an all-out assault on academic
freedom, critical scholarship, and the very idea of the university as a
place to question and think.[43]

In a similar manner, any academic and scientific knowledge that
challenged the rational foundations of these antidemocratic worldviews was either erased from government policies or attacked by
government talking heads as morally illegitimate, politically offensive, or in violation of patriotic correctness. Scientists who resisted
the ban on stem cell research as well as the official government position on global warming, HIV transmission, and sex education were
intimidated by congressional committees, which audited their work
or threatened "to withdraw federal grant support for projects whose
content they find substantively offensive."[44] Educators who argued
for theoretical and policy alternatives to abstinence as a mode of sex
education were attacked, fired, or cut out of funding programs for
education. And when the forces of patriotic correctness joined the
ranks of market fundamentalists, higher education was increasingly
defined through the political lens of an audit culture that organized
learning around measurable outcomes rather than modes of critical
thinking and inquiry.

In the war being waged by right-wing extremists in order to divest
the university of its critical intellectuals and critically oriented curricula, programs, and departments, ACTA produced a booklet titled
*How Many Ward Churchills?* in which it insisted that the space that
separated most faculty from political radicals like Ward Churchill (con-
troversially fired by the University of Colorado in 2007—a decision
reversed by the courts in 2009) was small indeed, and that by protecting such individuals colleges and universities now "risk losing their
independence and the privilege they have traditionally enjoyed."[45]
And how do we know that higher education has fallen into such
dire straits? These apocalyptic conditions were revealed through an
inane summary of various course syllabi offered by respected universities that allegedly proved "professors are using their classrooms
to push political agendas in the name of teaching students to think
critically."[46] Courses that included discussions of race, social justice,
gender equality, and whiteness as a tool of exclusion were dismissed
as distorting American history, by which ACTA meant consensus history, a position made famous by the tireless efforts of Lynne Cheney,
who has repeatedly asserted that American history should be celebra-
tory even if it means overlooking "internal conflicts and the non-white

population."[47] Rather than discuss the moral principles or pedagogical values of courses organized around the need to address human suffering, violence, and social injustice, the ACTA report claimed that "[a]nger and blame are central components of the pedagogy of social justice."[48] In the end, the listing of course descriptions was designed to alert administrators, governing boards, trustees, and tenure and hiring committees of the need to police instructors in the name of "impartiality." Presenting itself as a defender of academic freedom, ACTA actually wants to monitor and police the academy, just as Homeland Security monitors the reading habits of library patrons and the National Security Agency spies on American citizens without first obtaining warrants.

Despite its rhetoric, ACTA is not a friend of the principle of academic freedom or diversity. Nor is it comfortable with John Dewey's insistence that education should be responsive to the deepest conflicts of our time. And while the tactics to undermine academic freedom and critical education have grown more sophisticated, right-wing representations of the academy have become more shrill. For instance, James Pierson in the conservative *Weekly Standard* claimed that when 16 million students enter what he calls the "left-wing university," they will discover that "[t]he ideology of the left university is both anti-American and anticapitalist."[49] And for Roger Kimball, editor of the conservative journal *The New Criterion*, the university has been "corrupted by the values of Woodstock...that permeate our lives like a corrosive fog." He asks, "Why should parents fund the moral decivilization of their children at the hands of tenured antinomians?"[50] While relying on the objectification of youth, such anti-intellectualism reveals little understanding of how it does a disservice to young people, who have historically represented insightful and challenging views of social issues. Another example of these distortions occurred when former Republican presidential candidate Reverend Pat Robertson proclaimed that there were at least "thirty to forty thousand" left-wing professors or, as he called them, "termites that have worked into the woodwork of our academic society.... They are racists, murderers, sexual deviants and supporters of al-Qaeda—and they could be teaching your kids! These guys are out and out communists, they are propagandists of the first order. You don't want your child to be brainwashed by these radicals, you just don't want it to happen. Not only be brainwashed but beat up, they beat these people up, cower them into submission."[51] Robertson's comments mask a fundamental fear of young people in the guise of protecting them. The teachers or institutions do not pose nearly as much of a risk to Robertson's worldview as the young people themselves—those who could possibly go out into the world and actively try to change it. Most right-wing ideologues are more subtle and more insidious than Robertson, having dressed up their rhetoric in the language of fairness and balance, thereby cleverly expropriating, as Jonathan Cole suggests, "key terms in the liberal lexicon, as if they were the only true champions of freedom and diversity on campuses."[52] Inflated rhetoric aside, the irony of such rallying cries against "liberal propaganda" is that they support a conservative project designed to impose more oversight and control of the university, discriminate against liberal students and faculty, legislate more outside control over teacher authority, enact laws to protect conservative students from pedagogical "harassment" (that is, views differing from their own), and pass legislation that regulates the hiring process.

As I have pointed out in *The University in Chains*, one of the most powerful and well-known spokespersons leading the effort for "academic balance" is David Horowitz, president of the Center for the Study of Popular Culture and the ideological force behind the online magazine *FrontPageMag.com*. A self-identified former left-wing radical who has since become a right-wing conservative, he is the author of over 20 books and founder of Students for Academic Freedom, a national watchdog group that monitors what professors say in their classrooms. He is also the creator of *DiscovertheNetworks.org*, an online database whose purpose is to "catalogue all the organizations and individuals that make up" what he loosely defines in sweeping monolithic terms as "the Left."[53] As one of the most forceful voices in the assault on higher education, Horowitz has used the appeal to intellectual diversity and academic freedom with great success to promote his Academic Bill of Rights (ABOR),[54] the central purpose of which, according to Horowitz, is "to enumerate the rights of students to not be indoctrinated or otherwise assaulted by political propagandists in the classroom or any educational setting."[55] This rhetoric of student rights, allegedly defending youth, actually destroys students' access to a range of ideas, including the ones most prevalent among established scholars and validated by rigorous peer-review processes. Horowitz's case for the Academic Bill of Rights rests on a series of faulty empirical studies, many conducted by right-wing associations, which suggest left-wing views completely dominate the academy.[56] The studies look compelling until they are more closely examined.[57] For example, they rarely look at colleges, departments, or programs outside of the social sciences and humanities, thus excluding a large portion of the campus. According to the *Princeton Review*, four of the top-ten most popular subjects are business administration and management, biology, nursing, and computer science, none of which is included in Horowitz's data.[58] While it is very difficult to provide adequate statistics regarding the proportion of liberals to conservatives in academe, a University of California at Los Angeles report surveyed over 55,000 full-time faculty and administrators in 2002–2003 and found that "48 percent identified themselves as either liberal or far left; 34 percent as middle of the road, and ... 18 percent as conservative or far right."[59] All in all, 52.3 percent of college faculty either considered themselves centrist or conservative, suggesting that balance is far less elusive than Horowitz would have us believe. Furthermore, a 2006 study by the journal *Public Opinion Quarterly* argues that "recent trends suggest increased movement to the center, toward a more moderate faculty."[60] But there is more at stake here than the reliability of statistical studies measuring the voting patterns, values, and political positions of faculty. There is also the issue of whether such studies tell us anything at all about what happens in college classrooms. What correlation is to be correctly assumed between a professor's voting patterns and how he or she teaches a class? Actually, none. How might such studies deal with people whose political positions are not so clear, as when an individual is socially conservative but economically radical? And are we to assume that there is a correlation between "one's ideological orientation and the quality of one's academic work"?[61]

Then, of course, there are the questions that the right-wing commissars refuse to acknowledge: Who is going to monitor and determine what the politics of potential new hires, existing faculty members, and departments should be? How does such a crude notion of politics mediate disciplinary wars between, for instance, those whose work is empirically driven and those who adhere to qualitative methods? And if balance implies that all positions are equal and deserve equal time in order not to appear biased, should universities give equal time to Holocaust deniers, to work that supported apartheid in South Africa, or to proslavery advocates, to name but a few? Moreover, as Russell Jacoby points out with a degree of irony, if political balance is so important, then why isn't it invoked in other commanding sectors of society such as the police force, Pentagon, FBI, and CIA?[62]

The right-wing demand for balance also deploys the idea that conservative students are relentlessly harassed, intimidated, or unfairly graded because of their political views, despite their growing presence on college campuses and the generous financial support they receive from over a dozen conservative institutions. One place where such examples of alleged discrimination can be found is on the Web site of Horowitz's Students for Academic Freedom (SAF), whose credo is "You can't get a good education if they're only telling you half the story."[63] SAF has chapters on 150 campuses and maintains a Web site where students can register complaints. Most complaints express dissatisfaction with teacher comments or assigned readings that have a left-liberal orientation. Students complain, for instance, about reading lists that include books by Howard Zinn, Cornel West, or Barbara Ehrenreich. Others protest classroom screenings of Michael Moore's *Fahrenheit 9/11* or other documentary films such as *Super Size Me* and *Wal-Mart: The High Cost of Low Price*. Here is one student's complaint: "This class was terrible. We were assigned 3 books, plus a course reader! I don't think that just because a professor thinks they have the right to assign anything they want that they should be able to force us to read so much. In fact, I think the professor found out my religious and political beliefs and this is why he assigned so much reading."[64] Another student felt harassed because she had to read a text in class titled *Fast Food Nation*, which is faulted for arguing in favor of government regulation of the food industry. This is labeled "left indoctrination."[65]

What is especially disturbing about these complaints is that aggrieved students and their sympathizers appear entirely indifferent to the degree to which they not only enact a political intrusion into the classroom but also undermine the concept of education and professional academic standards that provide the basis for what is taught in classrooms, the approval of courses, and who is hired to teach such courses. Education is about fostering the conditions in which youth can make up their own minds, not indoctrinated. Horowitz's view of education as a one-way, top-down learning process is utterly facile, although it is telling: conservatives are most comfortable with precisely this kind of hierarchical authority structure and would like to see it emulated in the classroom. The complaints by conservative students often share the premise that because they are "consumers" of education, they have a right to demand what should be taught, as if knowledge is simply a commodity to be purchased according to one's taste. Awareness of academic procedures, research assessed by peer review, and basic standards for reasoning, as well as an understanding that professors earn a certain amount of authority because they are familiar with a research tradition and its methodologies, significant scholarship, and history, is entirely absent from such complaints that presuppose students have the right to listen only to ideas they agree with and to select their own classroom reading materials. Because some students disagree with an unsettling idea does not mean that they should have the authority, expertise, education, or power to dictate for all their classmates what should be stated, discussed, or taught in a classroom. What is lost in these arguments is the central pedagogical assumption that teaching is about activating and questioning all forms of knowledge, providing young people with the tools to critically engage what they know and to recognize the limits of their own knowledge. It is also about learning to think from the place of the other, to "raise one's self-reflexiveness to the highest maximum point of intensity."[66]

Defending higher education from this brand of anti-intellectualism is not motivated by "political bias" on the part of so-called left-wing universities. It is motivated, quite simply, by a principle informing all academic inquiry and education: intellectual responsibility involves an ongoing search for knowledge that enables a deeper and better understanding of the world. It is on these grounds that higher education must be defended. Neither academics nor students can ignore the democratic principles and conditions that make such knowledge available or even possible, that is, the conditions that enable critical scholarship and critical pedagogy both to survive and to flourish. Critical pedagogy is about teaching students how to hold authority and power accountable, providing them with the tools to make judgments freed from "the hierarchies of [official] knowledge" that attempt to shut down critical engagement. Such pedagogical tools are necessary for what Jacques Rancière calls "dissensus" or taking up a critical position that challenges the dogma of common sense.[67] As he puts it, "the work of dissensus is to always reexamine the boundaries between what is supposed to be normal and what is supposed to be subversive, between what is supposed to be active, and therefore political, and what is supposed to be passive or distant, and therefore apolitical."[68] Dissensus does more than call for "a modification of the sensible";[69] it also demands a utopian pedagogy that "provides names that one can give to ... the landscape of the possible," a landscape in which there is no room for the "machine that makes the 'state of things' unquestionable" and that insists upon a "declaration of our powerlessness."[70] In this way, critical pedagogy is about providing the conditions for students to be agents in a world that needs to be interrogated as part of a broader project of connecting the search for knowledge, truth, and justice to the ongoing tasks of democratizing both the university and larger society.

For many conservatives, the commitment to critical thinking and self-governance and the notion of pedagogy as a political and moral practice rather than as a disinterested technical task are simply outcomes of political indoctrination. Their attack on the university betrays a lack of trust in youth and a desire to retain power and authority in the hands of an unaccountable elite. For instance, Horowitz advocates in his book *The Professors* for a system of higher education that effectively depoliticizes pedagogy, deskills faculty, and infantilizes students, and he supports this position through the charge that a number of reputable scholars who take matters of critical thinking seriously are in reality indoctrinating their students with their own political views.[71] The book, as detailed by a report of the Free Exchange on Campus organization, is an appalling mix of falsehoods, lies, misrepresentations, and unsubstantiated anecdotes.[72] Not only does Horowitz fail to include in his list of "dangerous" professors one conservative academic, but many professors are condemned simply for what they teach, as Horowitz actually has little or no ammunition against *how* they teach. For example, Professor Lewis Gordon is criticized for including "contributions from Africana and Eastern thought" in his course on existentialism.[73] This is an utterly baffling criticism since Lewis Gordon is one of the world's leading African existential philosophers, a philosopher, moreover, who recognizes that "the body of literature that constitutes European existentialism is but one continent's response to a set of problems that date from the moment human beings faced problems of anguish and despair."[74] Horowitz's endless invective against critical intellectuals, all of whom he seems to consider left-wing, is perfectly captured in a comment he made on Dr. Laura's talk show in which he told the listening audience that "campus leftists hate America more than the terrorists."[75] This kind of diatribe has more in common with Sarah Palin's fearmongering remarks in the 2008 presidential campaign than it does with engaging in serious modes of analysis.

How does one take seriously Horowitz's call for fairness when he labels the American Library Association in his online magazine as "a terrorist sanctuary,"[76] or describes Noam Chomsky, whom the *New Yorker* named "one of the greatest minds of the 20th century,"[77] as "demonic and seditious" and claims the purpose of Chomsky's work is "to incite believers to provide aid and comfort to the enemies of the U.S."?[78] Indeed, what is one to make of Horowitz's online "A Guide to the Political Left" in which the mild-mannered film critic Roger Ebert occupies the same ideological ground as Omar Abdel Rahman, the mastermind of the 1993 World Trade Center bombing? Can one really believe that Horowitz is a voice for unbiased and open inquiry when he portrays as activists for "left-wing agendas and causes" the late Peter Jennings, Supreme Court Justice Ruth B. Ginsburg,

Garrison Keillor, and Katie Couric?[79] But apparently politicians at all levels of government *do* take Horowitz seriously. In 2005, Florida legislators considered a bill inspired by the ABOR that would provide students with the right to sue their professors if they felt their views, such as a belief in Intelligent Design, were disrespected in class.[80] At the federal level, the ABOR legislation made its way through various House and Senate Committees with the firm backing of a number of politicians and was passed in the House of Representatives in March 2006, but went no further.[81] In 2007, a Senate committee in Arizona passed a bill in which faculty could be fined up to $500 for "advocating one side of a social, political, or cultural issue that is a matter of partisan controversy."[82]

As Stanley Fish has argued, "balance" is a flawed concept and should be understood as a political tactic rather than an academic value.[83] The appeal to balance is designed to do more than get conservatives teaching in English departments, promote intellectual diversity, or protect conservative students from the supposed horrors of left-wing indoctrination; its deeper purpose is to monitor pedagogical exchange through government intervention, calling into question the viability of academic integrity and undermining the university as a public sphere that educates students as critically engaged and responsible citizens in the larger global context. The attack by Horowitz and his allies against liberal faculty and programs in the social sciences and humanities such as Middle Eastern studies, women's studies, and peace studies has opened the door to a whole new level of assault on academic freedom, teacher authority, and critical pedagogy.[84] These attacks, as I have pointed out, are much more widespread and, in my estimation, much more dangerous than the McCarthyite campaign several decades ago.

In response to this attack on academic freedom, unfortunately even the most spirited defenders of the university as a democratic public sphere too often overlook the ominous threat being posed to what takes place in the classroom, and, by extension, to the very nature of pedagogy as a political, moral, and critical practice.[85] The concept of balance demeans teacher authority by suggesting that a political litmus test is the most appropriate consideration for teaching, and it devalues students by suggesting that they are happy robots, interested not in thinking but in merely acquiring skills for jobs. In this view, students are rendered incapable of thinking critically or engaging knowledge that unsettles their worldviews and are considered too weak to resist ideas that challenge their commonsense understanding of the world. And teachers are turned into instruments of official power and apologists for the existing order. Teacher authority can never be neutral; nor can it be assessed in terms that are narrowly ideological. It is always broadly political and interventionist in terms of the knowledge effects it produces, the classroom experiences it organizes, and the future it presupposes in the countless ways in which it addresses the world. Teacher authority suggests that as educators we must make a sincere effort to be self-reflective about the value-laden nature of our authority while rising to the fundamental challenge of educating students to take responsibility for the direction of society.

It should come as no surprise that many religious and political conservatives view critical pedagogy as dangerous, often treating it with utter disdain or contempt. Critical pedagogy's alleged crimes can be found in some of its most important presuppositions about the purpose of education and the responsibility of educators. These include its central tenet that at the very core of education is the task of educating students to become critical agents who actively question and negotiate the relationships between theory and practice, schooling and everyday life, and the larger society and the domain of common sense. At stake here is a notion of teaching that refuses simply to serve government power, national interests, a rigid social order, and officially sanctioned views of the world. Also at stake here is the recognition that critical pedagogy opens up a space where students should be able to come to terms with their own power as critical agents; that is, it provides a sphere where the unconditional freedom to question and take a stance is central to the purpose of the university and also to democracy itself.[86] In this discourse, pedagogy always represents a commitment to the future, and it remains the task of educators to point the way to a more socially just world, a world in which the discourses of critique and possibility in conjunction with the values of reason, freedom, and equality function to better, as part of a broader democratic project, the grounds upon which life is lived. This is not a prescription for political indoctrination; rather, it is a project that gives education its most valued purpose and meaning. In other words, critical pedagogy forges both critique and agency through a language of skepticism and possibility and a culture of openness, debate, and engagement among students and teachers—all elements that are now at risk in the latest and most dangerous attack on higher education. Not only is academic freedom defended in the justification for critical pedagogical work, but it is also importantly safeguarded through the modes of academic labor and governance that connect the search for knowledge with increasing the capacity for all members of society to acquire the freedom to learn and to engage in mutual criticism that is "based in the quality of their ideas, rather than in their social positions."[87]

While liberals, progressives, and left-oriented educators and youth have increasingly opposed the right-wing assault on higher education, they have not done enough either theoretically or politically. While there is a greater concern about the shameless state of non-tenured and part-time faculty in the United States (actually, an under-the-radar parallel alternative to the traditional tenure system), such concerns have not been connected to a full-spirited critique of other antidemocratic forces now affecting higher education through a growing managerial culture and a neoliberal approach to university governance.[88] Neoliberalism makes possible not only the ongoing corporatization of the university and the increasing militarization of knowledge but also the powerlessness of faculty, staff, and students who are increasingly treated by administrators as replaceable populations. It is well known that power relations within universities and colleges today are top-heavy, controlled by trustees and administrators and removed from the hands of those who actually do the work. Power has instead become centralized largely in the hands of administrators, who are close to business, industry, and the national security state. If it is going to have a future as a democratic public sphere, higher education must divorce itself from those knowledge forms, underlying values, practices, ideologies, social relations, and cultural representations associated with the intensification and expansion of corporate and military culture. With respect to the latter, it is clear that higher education has no legitimate or ethical reason for engaging in practices that are organized largely for the production of violence.

It is important to reclaim higher education as a site of moral and political practice whose purpose is not only to introduce students to the great reservoir of diverse intellectual ideas and traditions but also to engage those inherited bodies of knowledge thorough critical dialogue, analysis, and comprehension. As students increasingly find themselves part of an indentured generation, there is a need for educators and others to once again connect matters of equity and excellence as two inseparable freedoms. Students' right to access higher education, to participate in the governance of the university, and to freely express and debate their ideas in the classroom must be defended intellectually and financially. Unless parents, labor unions, students and concerned individuals mobilize to protect the institutionalized relationships between democracy and pedagogy, teacher authority and classroom autonomy, higher education will be at the mercy of a right-wing revolution that views democracy as an excess and the university as a threat to society at large.

Pedagogy must be understood as central to any discourse about academic freedom, but, more important, it must be understood as one of the most crucial referents we have for understanding the politics of education and defending the university as one of the very few remaining democratic public spheres in the United States today. As Ian Angus rightly argues, "The justification for academic freedom lies in the activity of critical thinking"[89] and the pedagogical and political conditions necessary to protect it. I believe that too many notions of academic freedom are defined through a privatized notion of individual freedom, largely removed from the issue of collective democratic governance, which is the primary foundation enabling academic freedom to become a reality. Right-wing notions of teaching and learning constitute a kind of anti-pedagogy, substituting conformity for dialogue and ideological inflexibility for critical engagement. Such attacks should be named for what they are—an affirmation of thoughtlessness, a disservice to young peoples' ability to question and be self-directed, and an antidote to the difficult process of self- and social criticism.[90] In spite of what conservatives claim, this type of pedagogy is not education, but a kind of training that produces a flight from self and society. Its outcome is not a student who feels a responsibility to others, but one who feels the presence of difference as an unbearable burden to be contained or expelled. In this way, it becomes apparent that the current right-wing assault on higher education is directed not only against the conditions that make critical pedagogy possible but also against the possibility of raising questions about the real problems facing higher education and youth today, who should be given opportunities to engage knowledge critically, to make judgments, to intervene in the world, and to assume responsibility for what it means to know something.

Higher education is increasingly becoming unaffordable for all but the most prosperous of students. At its best, higher education should be free for all students simply because it is not an entitlement but a right, one that is crucial for a functioning democracy. Hence, the call for strategies to retake higher education also argues for making higher education available to everyone, regardless of wealth and privilege. Higher education has to be democratized and cannot be tuition-driven, a trend that reinforces differential opportunities for students based on their ability to pay. At the very least, student loans must be replaced with a combination of outright financial grants and work-study programs, thus making it possible for all individuals who want to obtain higher education, and especially for those marginalized by class and race, to be able to do so. Moreover, making higher education free would eliminate the need for those who cannot afford higher education to volunteer to serve in the military and put their lives in danger in order "to gain the educational opportunities that arguably would be the right of every citizen in a less shameless democracy."[91]

The ongoing vocationalization of higher education, the instrumentalization of the curriculum, the increasing connection between the military and universities through joint research projects and Pentagon scholarships, and the transformation of students into consumers have undermined colleges and universities in their efforts to offer students the knowledge and skills they need for learning how to govern as well as for developing the capacities necessary for deliberation, reasoned argumentation, and the obligations of civic responsibility. Higher education has become part of a market-driven and militarized culture, imposing upon academics and students new modes of discipline that close down the spaces to think critically, undermine substantive dialogue, and restrict students from thinking outside of established expectations. The conservative pedagogical project, despite paying lip service to the idea of "balance," is less about promoting intellectual curiosity, understanding the world differently, or enabling students to raise fundamental questions about "what sort of world one is constructing."[92] On the contrary, its primary purpose is to produce dutiful subjects willing to sacrifice their sense of agency for a militaristic sense of order and unquestioning respect for authority. All this leads toward a society in which there is no end to the increasing role of part-time labor, the commodification of knowledge, the rise of an expanding national security state, the hijacking of public spheres by corporate and militarized interests, and the increasing attempts by right-wing extremists to turn education into job training and public pedagogy into an extended exercise in patriotic xenophobia. This is more than a pedagogy for conformity: it is also a recipe for a type of thoughtlessness that, as Hannah Arendt reminds us, is at the heart of totalitarian regimes.[93]

In light of this right-wing assault on critical thought and youth, educators have a political and moral responsibility to critique the university as a major element in the military-industrial-academic complex. At the very least, this means being attentive to the ways in which conservative pedagogical practices deny the democratic purposes of education and the role of young people in fostering democracy, and so undermine the possibility of a critical citizenry. Yet such a critique, while important, is not enough. Academics also have a responsibility to make clear higher education's association with other memories, brought back to life in the 1960s, in which the academy was remembered for its "public role in developing citizenship and social awareness—a role that shaped and overrode its economic function."[94] Such memories, however uncomfortable to the new corporate managers of higher education, must be nurtured and developed in defense of higher education as an important site of both critical thought and democratization. Instead of a narrative of decline, young people need a discourse of critique and resistance, possibility and hope. Such memories both recall and seek to reclaim how consciousness of the public and democratic role of higher education, however imperfect, gives new meaning to its purpose and raises fundamental questions about how knowledge can be emancipatory and how an education for democracy can be both desirable and possible.

What needs to be understood is that higher education may be one of the few public spheres left where knowledge, values, and learning offer a glimpse of the promise of education for nurturing critical hope and a substantive democracy.[95] It may be the case that everyday life is increasingly organized around market principles, but confusing democracy with market relations hollows out the legacy of higher education, whose deepest roots are moral, not commercial. In defending young people's ability to access and to learn from educational rather than corporate institutions, we must heed the important insight expressed by Federico Mayor, the former director general of UNESCO, who insists that "[y]ou cannot expect anything from uneducated citizens except unstable democracy,"[96] or, what is becoming increasingly apparent, something even worse. As the free circulation of ideas is replaced by ideas managed and disseminated by the corporate media, ideas become banal, if not reactionary; intellectuals who engage in dissent are viewed or dismissed as either irrelevant, extremist, or un-American; and complicit public relations intellectuals dominate the media, all too willing to internalize co-optation and reap the rewards of venting insults at their alleged opponents. What is lost in these antidemocratic practices are the economic, political, educational, and social conditions that provide a supportive culture for democracy to flourish. This is, in part, a deeply pedagogical and educational issue that should not be lost on either intellectuals or those concerned about the purpose and meaning of higher education and youth. Only through such a supportive and critical educational culture can students learn how to become individual and social agents—rather than merely disengaged spectators—willing not only to think otherwise but also to act upon civic commitments that "necessitate a reordering of basic power arrangements" fundamental to promoting the common good and producing a meaningful democracy.[97]

The current right-wing assault on higher education is in reality an attack on the most rudimentary conditions of democratic politics. Democracy cannot work if citizens are not autonomous, self-judging, curious, reflective, and independent—qualities that are indispensable for students if they are going to make vital judgments and choices about participating in and shaping decisions that affect everyday life, institutional reform, and governmental policy in their own country and around the globe. This means educators both in and outside of the university need to reassert pedagogy as the cornerstone of democracy by demonstrating in our classrooms and also to the broader public that it provides the very foundation for students to learn not merely how to be governed but also how to be capable of governing. What is even more crucial, as Stuart Hall points out, is the urgent need for educators to provide students with "[c]ritical knowledge [that is] *ahead* of traditional knowledge . . . *better* than anything that traditional knowledge can produce, because only serious ideas are going to stand up." At the same time, there is also the need to recognize "the social limits of academic knowledge. Critical intellectual work cannot be limited to the university but must constantly look for ways of making that knowledge available to wider social forces."[98] If Hall is right, and I think he is, educators have a pedagogical responsibility to make knowledge meaningful in order to make it critical and transformative. Such knowledge would expand the range of human possibilities by connecting what young people know and how they come to know to instilling in them both "a disgust for all forms of socially produced injustice"[99] and the desire to make the world different from what it is.

## ACADEMICS AND PUBLIC LIFE

Addressing education as a democratic endeavor begins with the recognition that higher education is more than an investment opportunity; citizenship is more than conspicuous consumption; learning is more than preparing students for the workplace, however important that task might be; and democracy is more than making choices at the local mall. If higher education is to reclaim itself as a site of critical thinking, collective work, and public service, educators and students will have to redefine the knowledge, skills, research, and intellectual practices currently favored in the university. Central to such a challenge is the need to position intellectual practice "as part of an intricate web of morality, rigor, and responsibility" that enables academics to speak with conviction, use the public sphere to address important social problems, and demonstrate alternative models for bridging the gap between higher education and the broader society.[100] Connective practices are key: it is crucial to develop intellectual practices that are collegial rather than competitive, to refuse the instrumentality and privileged isolation of the academy, to link critical thought to a profound impatience with the status quo, and to connect human agency to the idea of social responsibility and the politics of possibility.

Connection also means being openly and deliberately critical and worldly in one's intellectual work. Increasingly, as universities are shaped by a culture of fear in which dissent is equated with treason, the call to be objective and impartial, whatever one's intentions, can easily echo what George Orwell called the "official truth" or the establishment point of view. Lacking a self-consciously democratic political focus, teachers and students are often reduced to the role of a technician or functionary engaged in formalistic rituals, unconcerned with the disturbing and urgent problems that confront the larger society or the consequences of one's pedagogical practices and research undertakings. In opposition to this model, with its claims to and conceit of political neutrality, I argue that academics should combine the mutually interdependent roles of critical educator and active citizen. This requires finding ways to connect the practice of classroom teaching with the operation of power in the larger society and to provide the conditions for students to view themselves as critical agents capable of making those who exercise authority and power accountable.

Education cannot be divorced from democracy; and as such, it must be understood as a deliberately informed and purposeful political and moral practice, as opposed to one that is either doctrinaire or instrumentalized, or both. In a society that remains troublingly resistant to or incapable of questioning itself, one that celebrates the consumer over the citizen and willingly endorses the narrow values and interests of corporate power, the importance of the university as a place of critical learning, thoughtfulness, moral responsibility, and social justice advocacy becomes all the more imperative. Moreover, the distinctive role that faculty play in this ongoing pedagogical project of democratization and learning, along with support for the institutional conditions and relations of power that make it possible, must be defended as part of a broader discourse of excellence, equity, and democracy. As Sheldon Wolin points out, "For its part, democracy is ultimately dependent on the quality and accessibility of public education, especially of public universities. Education per se is not a source of *democratic* legitimacy: it does not serve as a justification for political authority, yet it is essential to the practice of citizenship."[101]

For education to be civic, critical, and democratic rather than privatized, militarized, and commodified, the work that academics do cannot be defended exclusively within the discourse of specialization, technological mastery, or a market-driven rationality concerned about efficiency and profit margins. On the contrary, academic labor is distinctive by virtue of its commitment to modes of education that take seriously John Dewey's notion that democracy is a "way of life" that must be constantly nurtured and defended, or as Richard Bernstein puts it:

Democracy, according to Dewey, does not consist exclusively of a set of institutions, formal voting procedures, or even legal guarantee of rights. These are important, but they require a culture of everyday democratic cooperative *practices* to give them life and meaning. Otherwise institutions and procedures are in danger of becoming hollow and meaningless. Democracy is "a way of life," an ethical ideal that demands *active* and *constant* attention. And if we fail to work at creating and re-creating democracy, there is no guarantee that it will survive. Democracy involves a reflective faith in the capacity of all human beings for intelligent judgment, deliberation, and action if the proper social, educational, and economic conditions are furnished.[102]

Education should not be decoupled from what Jacques Derrida calls a democracy to come, that is, a democracy that must always "be open to the possibility of being contested, of contesting itself, of criticizing and indefinitely improving itself."[103] Democracy is not cheap and neither are the political, economic, and social conditions that make it possible. If academics believe that the university is a space for and about democracy, they need to become more attentive to addressing the racial, economic, and political conditions that fill their ranks with adjuncts, remove faculty from exercising power in university governance, and work towards eliminating the economic conditions that prevent working-class and middle-class youth from getting a decent post-secondary education.

Moreover, a critical pedagogy that values a democratic and open society should be engaged at all levels of schooling. It must gain part of its momentum in higher education among students who will go back to the schools, churches, synagogues, and workplaces in order to produce new ideas, concepts, and critical ways of understanding the world in which young people and adults live. This is a notion of intellectual practice and responsibility that refuses the insular, overly pragmatic, and privileged isolation of the academy while affirming a broader vision of learning that links knowledge to the power of self-definition and to the capacities of students to expand the scope of democratic freedoms, particularly those that address the crisis of education, politics, and the social as part and parcel of the crisis of democracy itself. This is the kind of intellectual practice that Zygmunt Bauman calls "taking responsibility for our responsibility,"[104] one that is attentive to the suffering of others and "will not allow conscience to look away or fall asleep."[105]

In order for pedagogy that encourages critical thought to have a real effect, it must include the message that all citizens, old and young, are equally entitled, if not equally empowered, to shape the society in which they live. If educators are to function as public intellectuals, they need to provide the opportunities for students to learn that the relationship between knowledge and power can be emancipatory, that their histories and experiences matter, and that what they say and do counts in their struggle to unlearn dominating privileges, productively reconstruct their relations with others, and transform, when necessary, the world around them. Simply put, educators need to argue for forms of pedagogy that close the gap between the university and everyday life. Their curricula need to be organized around knowledge about communities, cultures, and traditions that give students a sense of history, identity, and place. Said illuminates this process when he urges academics and students to accept the demands of "worldliness," which include "lifting complex ideas into the public space," recognizing human injury inside and outside of the academy, and using theory as a critical resource to change things.[106] Worldliness suggests that we must not be afraid of controversy and that we must make connections that are otherwise hidden, deflate the claims of triumphalism, and bridge intellectual work and the operation of politics. It means combining rigor and clarity, on the one hand, and civic courage and political commitment, on the other.

A critically engaged pedagogy also necessitates that we incorporate in our classrooms those electronically mediated knowledge forms that constitute the terrain of mass and popular culture. I am referring here to the world of media texts—videos, films, the Internet, podcasts, and other elements of the new electronic technologies that operate through a combination of visual and print culture. Such an approach not only challenges the traditional definition of schooling as the only site of pedagogy by widening the application and sites of education to a variety of cultural locations but also alerts students to the educational force of the culture at large, what I have called elsewhere the field of public pedagogy.

Any viable notion of critical pedagogy should affirm and enrich the meaning, language, and knowledge forms that students actually use to negotiate and inform their lives. Academics can, in part, exercise their role as public intellectuals via such approaches by giving students the opportunity to understand how power is organized through an enormous number of "popular" cultural spheres, including libraries, movie theaters, schools, and high-tech media conglomerates that circulate signs and meanings through newspapers, magazines, advertisements, new information technologies, computers, and television programs. Needless to say, this position challenges neoconservative Roger Kimball's claim that "[p]opular culture is a tradition essential to uneducated Americans."[107] By laying claim to popular, mass, and alternative cultural spaces as important sites of public pedagogy, educators have the opportunity, if not the responsibility, to raise important questions about how knowledge is produced, circulated, and taken up in different pedagogical sites. They can also provide the foundation for students to become competent and critically versed in a variety of literacies (not just the literacy of print), while at the same time expanding the conditions and options for the roles students might play as cultural producers (as opposed to simply teaching them to be critical readers). At stake here is an understanding of literacy as both a set of competencies to be learned and a crucial condition for developing ways of intervening in the world.

I have suggested that educators need to become provocateurs; they need to take a stand while refusing to be involved in either a cynical relativism or doctrinaire politics. This suggests that central to intellectual life is the pedagogical and political imperative that academics engage in rigorous social criticism while becoming a stubborn force for challenging false prophets, fighting against the imposed silence of normalized power, and critically engaging all those social relations that promote material and symbolic violence.[108] There is a lot of talk among social theorists about the death of politics brought on by a negative globalization characterized by markets without frontiers, deregulation, militarism, and armed violence, all of which not only feed each other but produce global unlawfulness and reduce politics to merely an extension of war.[109] I would hope that, of all groups, educators would vocally and tirelessly challenge this ideology by making it clear that expanding the public good and promoting democratic social change are at the very heart of critical education and are pre-conditions for global justice. The potential for a better future further increases when critical education is directed toward young people. As a result, public and higher education may be among the few spheres left in which the promise of youth can be linked to the promise of democracy.

As the dark times that characterized the Bush years have come to an end and the promise of a more progressive model of governance and respect for education seems possible under the presidency of Barack Obama, it is worth remembering that higher education, even in its crippled state, still poses a threat to the enemies of democracy; it holds the promise, if rarely realized, of being able to offer students the knowledge and skills that enable them not only to mediate critically between democratic values and the demands of corporate power and the national security state but also to distinguish between identities founded on democratic principles, on the one hand, and subject positions steeped in forms of competitive, unbridled individualism that celebrate self-interest, profit-making, militarism, and greed, on the other. Education in this instance becomes both an ethical and a political referent; it furnishes an opportunity for adults to provide the conditions for young people to become critically engaged social agents. Similarly, it points to a future in which a critical education, in part, creates the conditions for each generation of youth to struggle anew to sustain the promise of a democracy that has no endpoint, but rather must be continuously expanded into a world of new possibilities and opportunities for keeping justice and hope alive.

I want to emphasize that how we view, represent, and treat young people should be part of a larger public dialogue about how to imagine a democratic future. Dietrich Bonhoeffer, the great Protestant theologian, believed that the ultimate test of morality resides in what a society does for its children. If we take this standard seriously, American society has deeply failed its children and its commitment to democracy. The culture of neoliberalism and consumer culture rest on the denial of both youth as a marker of the future and the social responsibility entailed by an acceptance of this principle. In other words, the current crisis of American democracy can be measured in part by the fact that too many young people are poor, lack decent housing and health care, and attend decrepit schools filled with overworked and underpaid teachers. These youth, by all standards, deserve more in a country that historically prided itself on its level of democracy, liberty, and alleged equality for all citizens. For many young people, the future looks bleak, filled with the promise of low-paying, low-skilled jobs, the collapse of the welfare state, and, if you are a person of color and poor, the threat of either unemployment or incarceration.

We have entered a period in which the war against youth, especially poor youth of color, offers no apologies because it is too arrogant and ruthless to imagine any resistance. But power as a form of domination is never absolute, and oppression always produces some form of resistance. For these reasons, the collective need and potential struggle for justice should never be underestimated even in the darkest of times. To confront the biopolitics of disposability and the war on young people, we need to create the conditions for multiple collective and global struggles that refuse to use politics as an act of war and markets as the measure of democracy. Fortunately, more and more young people nationally and internationally are mobilizing in order to fight a world dominated by corporate interests and are struggling to construct an alternative future in which their voices can be heard as part of a broader movement to make democracy and social justice realizable.

Education, when connected to social change, can help provide the knowledge, tools, and hope necessary to further motivate these young people, many of whom recognize that the world stands at a critical juncture and that they can play a crucial role in changing it. For many young people, social injustices that extend from class oppression to racial violence to the ongoing destruction of public life and the environment can no longer be tolerated. We have watched young people all over the globe march against the injustices of negative globalization in recent years. What needs to be stressed is that these are political and educational issues, not merely economic concerns.

Hannah Arendt insisted that making human beings superfluous is the essence of totalitarianism, and the war against youth and critical education suggests that a new form of authoritarianism is ready to take over if we cannot work together to develop a new politics, a new analytic of struggle, and, most importantly, a renewed sense of imagination, vision, and hope. The great abolitionist Frederick Douglass bravely argued that freedom is an empty abstraction if people fail to act, and "if there is no struggle, there is no progress."[110] We live in a historic moment of both crisis and possibility, one that presents educators, parents, artists, and others with the opportunity to take up the challenge of reimagining civic engagement and social transformation, but these activities have a chance of succeeding only if we also defend and reinvigorate the pedagogical conditions that enable the current generation of young people to nurture thoughtfulness, critical agency, compassion, and democracy itself. I realize this sounds a bit utopian, but we have few choices if we are going to fight for a future that enables young people to escape from a political order in which living either as a commodity or as part of the growing refuse of human disposability are the only choices through which they can make a claim on the future. Young people deserve more, and adults should embrace the responsibility to help make it happen.

# CHAPTER 4

# IN THE SHADOW OF THE GILDED AGE: BIOPOLITICS IN THE AGE OF DISPOSABILITY

The health of any given society can be understood through an examination of the attitudes, challenges, and realities that confront its youth on a daily basis. When young people in the United States are increasingly subject to forces that commodify them, criminalize them, and deem them unworthy of receiving a critical and laudable education, it bodes very ill for the nation as a whole. While it is important to explore the particular problems facing youth as a result of state and institutional policy and misrepresentation in the dominant media, this approach does not go far enough. What is emerging is a new global order in which the neoliberal logic of consuming and disposability reigns supreme, in spite of the current financial crisis. The issues of global democracy and universal access to quality education must be made central to any effort to address the plight of young people. At the same time, the issues facing youth are crucial to any conceptualization and future reality of global democracy. Young people—as a concrete embodiment and symptomatic reflection of the abstract forces that govern the social sphere—are one of the most significant modalities through which to understand and launch an effective resistance to neoliberalism as a political, economic, and social movement. Indeed, no rigorous attempt to examine the meaning, implications, and consequences of neoliberalism can do without a methodological approach that connects the particular, concrete realities of people's lives to the general phenomena governing the state and social structure, which although not visible to the eye, are no less real as they bear down on everyday existence.

This final chapter initiates a move from the particular focus of the first three chapters—the exploration of the problems facing young people in the United States—to a general exploration of several important features of neoliberal governance and the ideology of disposability, especially as they have manifested under the eight-year term of the George W. Bush administration. More importantly, it takes the specific elements of Bush's neoliberal agenda and contextualizes them within a range of philosophical positions produced in recent years in response to the emergence of neoliberalism on a global scale. Any effective resistance to the politics of disposability will need to address the full complexity and reach of neoliberalism at more than the levels of U.S. social policy: it will need to address the transformation of democracy worldwide as a way to conceive of politics in the twenty-first century. But to achieve such a view of the social totality requires a massive collaborative endeavor: this book's focus on youth offers one piece of a much larger puzzle, while the following chapter suggests a theoretical framework for how to integrate these various pieces into a whole. Much more work is waiting to be done. As a first step, however, I believe global resistance to neoliberalism must prioritize the safeguarding of youth, their education, and their rights, if the world has any chance to achieve sustainability—sustainability now being propounded as one of the central ideals of an emergent global public sphere. Conversely, the concept of youth has the potential to provide a universal value and ethical position that can unite people across the political spectrum and globally, but only if the particular contexts surrounding youth are combined with the larger contexts of national political culture and the expansion of neoliberalism across the globe.

## NEOLIBERALISM AND THE RETURN OF THE GILDED AGE

As the Bush administration neared the end of its political tenure, the *New York Times* ran an editorial on the last day of 2007 insisting that the United States had become unrecognizable as a democratic society. Declaring that "there are too many moments when we cannot recognize our country,"[1] the editorial enumerated a list of state-sanctioned abuses, including torture by the CIA and subsequent repeated violations of the Geneva Conventions, the web of legalized illegality enabling the Bush administration to spy on Americans, and the willingness of government officials to violate civil and constitutional rights without apology, all done under the aegis of conducting the war on terrorism. Steadfast in its condemnation of the Bush administration, the editorial board of the *New York Times* argued that the United States government had induced a "state of lawless behavior...since September 11, 2001."[2] The *New York Times* was not alone in its concern. The prominent writer Sidney Blumenthal, a former senior adviser to President Clinton, claimed that we now live under a government tantamount to "a national security state of torture, ghost detainees, secret prisons, renditions and domestic eavesdropping."[3] Bob Herbert, an op-ed writer for the *New York Times,* asserted that the dark landscapes of exclusion, secrecy, illegal surveillance, and torture produced under the Bush regime offered Americans nothing less than a "road map to totalitarianism."[4] The French philosopher Jacques Rancière, may be most concise in arguing that what we have witnessed during the last few decades, epitomized by the Bush administration, is an image of the future that exhibits a deep hatred of democracy.[5]

While there is little question that the new millennium witnessed the United States moving into lock-down (and lock-out) mode both at home and abroad—with its burgeoning police state, its infamous title as the world leader in jailing its own citizens, and its history of foreign and domestic "torture factories"[6]—it is a mistake to assume that the Bush administration is solely responsible for transforming the United States to the degree that it has now become unrecognizable to itself as a democratic nation. Such claims risk reducing the serious social ills now plaguing the United States to the reactionary policies of the Bush regime—a move that allows for complacency to set in as Bush's reign came to a close on January 20, 2009, and Barack Obama took over the White House. The complacency caused by the sense of regime change fails to offer a truly political response to the current crisis because it ignores the extent to which Bush's policies merely recapitulated Clinton-era social and economic policy. What the United States has become in the last decade suggests less of a rupture than an intensification of a number of already existing political, economic, and social forces that have unleashed the repressive antidemocratic tendencies lurking beneath the damaged heritage of democratic ideals. What marks the present state of American "democracy" is the uniquely bipolar nature of the degenerative assault on the body politic, which combines elements of unprecedented greed and fanatical capitalism, called by some the New Gilded Age,[7] with a daring kind of politics more ruthless and savage in its willingness to abandon—even vilify—those individuals and groups now rendered disposable within the "geographies of exclusion and landscapes of wealth"[8] that mark the new world order.

The first Gilded Age, occurring at the end of the nineteenth century, serves as both a historical landmark and a point of departure in American history. As a historical landmark, it marks the rise of the Robber Barons; the merging of various backlash, nativist, and right-wing populist movements; legally sanctioned segregationism; a celebration of free-market economics; evangelical revivals; law-and-order moralism; limited government; violent labor conflicts; massive inequality; and the rise of a daunting nationalist capitalist class.[9] As an all-embracing rationality, it made visible an economic, political, and cultural model that presented a powerful political challenge for various progressive struggles, which in turn needed to contest the official ideology, values, institutions, and social relations that violently ordered American society around the discourses of racism, greed, unencumbered individualism, and self-interest, while recasting the entirety of political, cultural, and social life in terms of the calculating logic of the market. Inherently antidemocratic, and steeped in glamour and violence, the Gilded Age eventually gave way to progressive movements committed to the strengthening of the social state and a renewed sense of social citizenship, initiated by President Franklin D. Roosevelt as the New Deal, whose social emphasis demonstrated both a conception of democracy—extended primarily to its white citizenry—that served as a partial corrective to the deprivation of the Great Depression of the 1930s and a political refusal to reproduce the corruption of turn-of-the century politics "with its minimal taxation, absence of regulation, and reliance on faith-based charity rather than government social programs."[10]

The Gilded Age offers us today a historical snapshot of the worst underpinnings of an unchecked and unregulated capitalism, state-sanctioned racial repression, and modes of subjectivity, ideology, and politics that undermine any vestige of moral and political values that could sustain the public good and nourish a flourishing democracy. With the Great Depression and the ensuing emergence of powerful labor unions, the establishment of the welfare state, and the redistribution of income and jobs, the worst excesses of the Gilded Age seemed to be under control, especially between the 1930s and the 1970s. Unfortunately, in the last few decades, the reformist legacy of the New Deal and its ideological successor, the Great Society, initiated by President Lyndon Johnson in the 1960s, have been removed from both the rhetoric of politics and the very meaning of governance. Fortunately, the Obama administration has gestured favorably towards this legacy of reform, though it remains to be seen if there will be any serious attempts to extend this historic tradition of substantive reform.

The new exorbitantly rich, along with conservative ideologues such as Rush Limbaugh and Marvin Olasky, now publicly preach and celebrate the gospel of wealth associated with that period in nineteenth-century American history when corporations ruled political, economic, and social life, and a divinely ordained entrepreneurial spirit brought great riches and prosperity to the rest of the country. Olasky, who was ironically the mastermind behind the term "compassionate conservativism," argues that one of the virtues of going back to the nineteenth century is that government did not make the mistake of providing public assistance to the poor, or, to put it more specifically, government did not commit the crime of helping the less fortunate.[11] Any talk about the period's excesses, cruelty, and injustices is now dismissed as left-wing propaganda.[12] Or it is buried conveniently within the discourse of denial. For instance, soon after Barack Obama's election, former United States senator, Phil Gramm, an arch advocate of free-market ideology, an aggressive proponent of deregulation while serving in the Senate, and a major force in creating the political conditions that created the financial and credit crisis facing the globe, claimed that the mortgage crisis was the result of " 'predatory borrowers' who took out mortgages they could not afford," rather than the predatory loan practices he endorsed and blocked legislation from curbing.[13] But Gramm was not only a poster boy for free-market ideology and deregulation, he was also well liked by the commercial banks and Wall Street, who donated generously to his political campaigns. Moreover, his market fundamentalism seemed entirely removed from any of the havoc it wreaked on the less fortunate; he argued over the years that "food stamps be cut because 'all of our people are fat,' [and it] was hard for him 'to feel sorry' for Social Security recipients and, as the economy soured . . . [he] called America 'a nation of whiners.' "[14]

Since the late 1970s, we have witnessed the return of the Gilded Age under the aegis of a new and more ruthless form of market fundamentalism that has been labeled neoliberalism.[15] As a political-economic-cultural project, neoliberalism functions as a regulative force, political rationale, and mode of governmentality. As a regulative force, neoliberalism organizes a range of flows, including people, capital, knowledge, and wealth, transforming relations between the state and the economy by renouncing "big government" (a code word for the social state) as wasteful and incompetent—except in the current financial crisis, when the bankers who have been living lives befitting Gilded Age excess are now relying on the support of the government to bail them out of financial debt. As the current financial meltdown makes clear, neoliberal ideology has played a powerful role in eliminating government regulation of corporate behavior, providing enormous tax breaks for the rich and for powerful corporations, pursuing free trade agreements, and privatizing government assets, goods, enterprises, and public responsibilities.[16]

Essential to neoliberalism's regulative policies and goals is transforming the social state into a corporate state, one that generously sells off public property to transnational corporations and awards military contracts to private defense contractors, and one that ultimately provides welfare to an opulent minority. Government activities and public goods are now given over to the private sphere. Corporations and religious organizations benefit from government largess, while any activity that might interfere with corporate power and profits is scrapped or dismantled, including environmental regulations, public education, and social welfare programs. Schools and libraries are now privatized; forests are turned over to logging companies; military operations are increasingly outsourced to private security firms like Blackwater, while private security services protect the gated communities of the rich; prisons are run as for-profit institutions by corporations; and public highways are managed and leased to private firms. Increasingly, government services are being sold to the highest bidder. In short, capital is now being redistributed upward, as power is being transferred from traditional political localities to transnational corporations whose influence and reach exceed the boundaries and constraints formerly regulated by the nation-state.

As a mode of rationality, neoliberalism enables and legitimates the practices of managerialism, deregulation, efficiency, cost-benefit analysis, expanding entrepreneurial forms, and privatization, all of which function in the interest of "extending and disseminating market values to all institutions and social action."[17] As Wendy Brown points out, under neoliberalism:

> [t]he political sphere, along with every other dimension of contemporary existence, is submitted to an economic rationality; or, put the other way around, not only is the human being configured exhaustively as *homo œconomicus*, but all dimensions of human life are cast in terms of a market rationality. While this entails submitting every action and policy to considerations of profitability, equally important is the production of all human and institutional action as rational entrepreneurial action, conducted according to a calculus of utility, benefit or satisfaction against a microeconomic grid of scarcity, supply and demand, and moral value-neutrality. Neoliberalism does not simply assume that all aspects of social, cultural, and political life can be reduced to such

a calculus; rather, it develops institutional practices and rewards for enacting this vision.[18]

Extending this mode of rationality, the neoliberal economy with its relentless pursuit of market values now encompasses the entirety of human relations. As markets are touted as the driving force of everyday life, big government is disparaged as inefficient, monopolistic, incompetent, and thus a threat to individual and entrepreneurial freedom, suggesting that power should reside in markets and corporations rather than in governments and citizens. Under neoliberal rationality, citizens assume the role of entrepreneurial actors, bonded investors, or avid consumers while the state promotes market values throughout every aspect of the social order. Rather than fade away as some proponents of globalization would have us believe, the state embraces neoliberal rationality as the regulating principle of society in that it no longer merely endorses market relations: it now must "think and behave like a market actor across all of its functions, including the law [just as] the health and growth of the economy is the basis of state legitimacy."[19] The social state now becomes the "market state" and the "state's relationship to its citizens resembles that between a corporation and consumers."[20] Under neoliberalism, everything is either made saleable or plundered for profit while every effort is made to reconstruct the predatory state at work prior to the New Deal. Public lands are looted by real estate developers and corporate ranchers. Government regulators look the other way as banks promote bad mortgages and loans, eventually forcing millions of people to lose their homes. Politicians willingly hand the public's airwaves over to broadcasters and large corporate interests without a dime going into the public trust. Within the corporate state, "a generalized calculation of cost and benefit becomes the measure of all state practices."[21] As the state openly embraces and responds to the demands of the market, it invites corporations to drive the nation's energy policies, and war industries are given the green light to engage in war profiteering as the government hands out numerous contracts without any competitive bidding. Similarly, political and natural disasters are turned into entrepreneurial opportunities, which mark the destruction of the social state, the sale of public infrastructures, the imposition of privatization schemes, and the privatization of the politics of governance.[22]

    As the axis of all social interaction, neoliberal rationality expands far beyond the operations of the corporate state, the production of goods, and the legislating of laws.[23] As a seductive mode of public pedagogy, neoliberalism extends and disseminates the logic of the market economy throughout society, shaping not only social relations, institutions, and policies but also desires, values, and identities in the interest of constructing "the citizen-subject of a neoliberal order."[24] Under neoliberal rationality and its pedagogical practices not only are the state and the public sectors beholden to the whims of market choices, but the citizen-subjects of such an order navigate the relationship between themselves and others around the calculating logics of competition, individual risks, self-interest, and a winner-take-all ethic reminiscent of the script played out daily on "reality television." Backstabbing, deception, and a childish hypermasculinity become the structuring ideals of a neoliberal pedagogy, endlessly reproduced in the dominant media as part of the broader project of banishing the "anxieties and complexities of moral choice" and individual conscience.[25] Moreover, the survivalist ethic of nineteenth-century social Darwinism has been invoked to reinforce notions of racial hierarchy, and the current neoliberal agenda has systematically sought to recreate racial segregation and exclusion through the restructuring of income policies.

Neoliberalism also connects power and knowledge to the technologies, strategies, tactics, and pedagogical practices key to the management and ordering of populations and to controlling consent. Michel Foucault's concept of governmentality is crucial for understanding not only how modes of thought, rationality, and persuasion are linked to technologies of governing but also how any understanding of government must consider the ways power works to create "the conditions of consensus or the prerequisites of acceptance."[26] As Thomas Lemke has pointed out, neoliberal modes of governmentality are important for developing the connection "between technologies of the self and technologies of domination, the constitution of the subject and the formation of the state."[27] As a powerful mode of public pedagogy, neoliberal ideology is located, produced, and disseminated from many institutional and cultural sites, ranging from the shrill noise of largely conservative talk radio to the halls of academia and the screen culture of popular media.[28] Mobilizing modes of official knowledge, mass-mediated desires, and strategies of power, these sites provide an indispensable political service in coupling "technologies of the self and [neoliberal] political rationalities"[29] as part of a broader effort to transform politics, restructure power relations, and produce an array of narratives and disciplinary measures.[30] As neoliberalism extends into all aspects of daily life, the boundaries of the cultural, economic, and political become porous and leak into each other, sharing the task, though in different ways, of producing identities, goods, knowledge, modes of communication, affective investments, and many other aspects of social life and the social order.[31]

Fundamental to the construction of the neoliberal subject is the acceptance of this official set of orthodoxies: the public sphere, if not the very notion of the social, is a pathology; consumerism is the most important obligation of citizenship; freedom is an utterly privatized affair that legitimates the primacy of property rights over public priorities; the social state is bad; all public difficulties are individually determined; and all social problems, now individualized, can be redressed by private solutions. The undermining of social solidarities and collective structures along with the collapsing of public issues into private concerns is one of the most damning elements of neoliberal rationality. Zygmunt Bauman elucidates this issue in the following comment: "In our 'society of individuals' all the messes into which one can get are assumed to be self-made and all the hot water into which one can fall is proclaimed to have been boiled by the hapless failures [of those] who have fallen into it. For the good and the bad that fill one's life a person has only himself or herself to thank or to blame. And the way the 'whole-life-story' is told raises this assumption to the rank of an axiom."[32] Once again, any notion of collective goals designed to deepen and expand the meaning of freedom and democracy as part of the vocabulary of the public good is derided as taxing-and-spending big government liberalism or simply dismissed as socialism—an argument that the Republican Party uses constantly to rebuff every element of the stimulus plans proposed by the Obama administration. More specifically, "[c]ollective goals such as redistribution, public health, and the wider public good have no place in this landscape of individual preferences."[33] Instead, neoliberal theory and practice give rise to the replacement of the social state with a market/punishing state in which political rights are strictly limited; economic rights are deregulated and privatized; and social rights are replaced by the call to individual preference schemes and self-reliance. Within the impoverished vocabulary of privatization, individualism, and excessive materialism that promises to maximize choice and to minimize taxation, the new citizen-consumer bids a hasty retreat from those public spheres that view critique as a democratic value, collective responsibility as fundamental to the nurturing of democracy, and the deepening and expanding of collective protections as a legitimate function of the state. Defined largely by "the exaggerated and quite irrational belief in the ability of markets to solve all problems,"[34] the public domain is emptied of the democratic ideals, discourses, and identities needed to address important considerations such as universal health care, ecologically responsible mass transit, affordable housing with ethical lending practices, subsidized care for the young and elderly, and government efforts to reduce carbon emissions and invest in new forms of energy. As safety nets and social services are being hollowed out and communities crumble and give way to individualized, one-man archipelagos, it is increasingly difficult to develop social movements that can act in concert to effect policies to meet the basic needs of citizens and to maintain the social investments needed to provide life-sustaining services.

In order to foreground the connection between the emergence of a neoliberal Gilded Age and what I call the "politics of disposability"—a politics in which matters of life, death, and survival become central—currently on display in U.S. policies at home and abroad,[35] I want to draw attention to yet another set of narratives operating in public life, different from the ones generally used to indict the authoritarian tendencies of the former Bush regime (for instance, the network of CIA-sponsored secret prisons, the undemocratic workings of an imperial presidency, the extralegal operations of power, the emergence of a security state in which every citizen is viewed as a potential terrorist, and the attitude that war is the only viable index of public policy, shared values, and political legitimacy). While the importance of recognizing and understanding such dangerous trends cannot be underestimated, there are other, less visible registers of democratic decline, consigned to the margins of the dominant media, that also signal the pervasive, predatory mode of politics, rationality, and domination that now characterizes everyday life in America and that needs to be addressed under the more progressive Obama administration.

## NARRATIVES OF THE NEW (OLD) GILDED AGE

For most members of the American public, discourses of barbarism are often projected elsewhere to distant places rather than acknowledged as present in their everyday experiences. But images of pathology and violence, stories of extreme cruelty, and jarring symbolic disruptions are no longer easily displaced onto the traces of a largely forgotten past or inscribed in pedagogies of remembrance associated with geographies of brutal ethnic and civil wars such as Rwanda, Darfur, or the "killing fields" of Cambodia. Today, a predatory mode of politics and its accompanying representations, images, and discourses are constitutive of how American society has increasingly come not only to privilege death over life but also to view death as a form of entertainment. One example of the predatory mindset entailed by neoliberal policies appeared in the *New York Times* at the beginning of 2008 and told the story of two elderly men who were arrested while "pushing a corpse, seated in an office chair, along the sidewalk to a check-cashing store to cash the dead man's Social Security check." In a desperate attempt to cash the late Mr. Cintron's $335 check, the two men "parked the chair with the corpse in front of Pay-O-Matic at 763 Ninth Avenue," a business that Mr. Cintron had frequented. The attempt failed, as the newspaper reported, because "[t]heir sidewalk procession had already attracted the stares of passers-by who were startled by the sight of the body flopping from side to side as the two men tried to prop it up, the police said."[36] Police and an ambulance arrived as the two men attempted to maneuver the corpse in the chair into the office. The story offered no reasons for such behavior and treated the narrative more as a kind of odd spectacle akin to the workings of the *Jerry Springer Show* than as a serious commentary about the sheer desperation that follows the collapse of the social state, accompanied by an ever-expanding poverty, volatility, and insecurity that encroach on whole populations in the United States.

Another even more brutal account in the mainstream press told of how a New York City police detective and his girlfriend kidnapped and forced a 13-year-old girl to provide sexual favors for the couple's friends and other interested buyers. According to the story, "the detective and his girlfriend would parade the girl at parties and other places where adult men had gathered and force her to have sex with them for money—$40 for oral sex, $80 for intercourse. The child was an investment. The couple allegedly told her that she had been purchased for $500—purchased, like the slaves of old, only this time for use as a prostitute."[37] While the story connected the fate of this young child to the growing sex trade in the United States, it said nothing about the ongoing reification of young girls in a market society that largely reduces them to commodities, sexual objects, and infantilized accessories for boys and men. While the sex trade clearly needs to be condemned and eliminated, it is an easy target politically and morally when compared to the music, advertising, television, and film industries that treat young people as merchandise, turn them into fodder for profit, and appear indifferent to the relentless public debasement of young girls and women.

A third story provides yet another glimpse of the treatment accorded to people deemed unworthy of humanity or dignity. In this narrative, Ben Zipperer contemplates the emergence of prison rodeos that are used to entertain large crowds by organizing games "where Americans buy tickets to watch inmates wrestle bulls and participate in crowd favorites like 'Convict Poker.' Also called 'Mexican Sweat,' the poker game consists of four prisoners who sit expectantly around a red card table. A 1,500-pound bull is unleashed, and the last convict to remain sitting wins. Especially thrilling for the audience is the chaotic finale 'Money the Hard Way' in which more than a dozen inmates scramble to snatch a poker chip dangling from the horns of another raging bull."[38]

In spite of their differences, all of these stories are bound together by a politics in which the logic of the marketplace is recalibrated to exploit society's most vulnerable—even to the point of transgressing the sanctity of the dead—and to inflict real horrors, enslavement, and injuries upon the lives of those who are poor, elderly, young, and disenfranchised, because they are without an economic role in the neoliberal order. And as the third story illustrates, a savage and fanatical capitalism offers a revealing snapshot of how violence against the incarcerated—largely black, often poor, and deemed utterly disposable—now enters the realm of popular culture by producing a type of racialized terrorism posing as extreme entertainment, while simultaneously recapitulating the legacy of barbarism associated with slavery.

Most of these stories place the blame for these crimes on individualized acts of cruelty and lawlessness. None offers a critical translation of the big picture, one that signals the weakening of social bonds and calls the very project of U.S. democracy into question. And yet these narratives demand something more, a different kind of optic capable of raising serious questions regarding the political culture and moral economy in which such representations are produced; the pedagogies of reification, vengeance, and sadistic pleasure that enable people to ignore their warning; and the inherent instability of a democracy that is willing to treat human beings as redundant and disposable, denied the rights and dignities accorded to the privileged economic subjects of the neoliberal order. And while such images conjure up startling representations of human poverty, misery, deepening inequality, and humiliation, they bear witness to a broader politics of exploitation and exclusion in which, as Naomi Klein points out, "Mass privatization and deregulation have bred armies of locked-out people, whose services are no longer needed, whose lifestyles are written off as 'backward,' whose basic needs go unmet."[39] These stories are decidedly selective, yet they point to something deeper still in the current mode of neoliberal regulation: the rise of a punishing state and its commitment to the criminalization of social problems, the unburdening of "human rights from a social economy,"[40] and the wide circulation of and pleasure in violent spectacles of insecurity and abject cruelty.

As the social state is displaced by the market, a new kind of politics is emerging in which some lives, if not whole groups, are seen as disposable and redundant. Within this new form of biopolitics—a political system actively involved in the management of the politics of life and death—new modes of individual and collective suffering emerge around the modalities and intersections of race and class. But what is important to recognize is that the configuration of politics that is emerging is about more than the processes of social exclusion or being left out of the benefits of the market: it is increasingly about a normalized and widely accepted reliance upon the alleged "invisible hand" of a market fundamentalism to mediate the most important decisions about life and death. In this case, the politics managing the crucial questions of life and death is governed by neoliberalism's power to define who matters and who doesn't, who lives and who dies. Questions about getting ahead no longer occupy a key role in everyday politics. For most people under the regime of neoliberalism, everyday life has taken an ominous turn and is largely organized around questions of who is going to survive and who is going to die. Under such circumstances, important decisions about life and death have given way to a range of antidemocratic forces that threaten the meaning and substance of democracy, politics, the human condition, and any viable and just vision of the future. In its updated version, neoliberal rationality also rules "our politics, our electoral systems, our universities, increasingly dominat[ing] almost everything, even moving into areas that were once prohibited by custom in our country, like commercializing childhood."[41]

In a society in which the public sphere is characterized by a culture of fear and public life has receded behind gated communities, a pervasive discourse of privatization coupled with the practice of brutalization embraces an utterly narrow and commodified definition of freedom and feeds a disinterest in politics, while closing down any sense of responsibility for those who in a neoliberal capitalist society represent the losers, the unemployed, the incarcerated, the poor, the young, and the elderly. Randy Martin captures the violence of this process in his comment: "Privatization, the state's internal war on behalf of a capital said to be able to manage itself, savages populations, subjecting private matters to the public violence of the market."[42] As the spaces where politics can occur are rendered as either commercial spheres, dizzying sources of financial gain, or advertisements for the profit-driven fantasies of the corporate elite, compassion turns to disdain for those who are considered without merit in a market economy and too poor to participate in the hyper-circuits of power that characterize the New Gilded Age. While predating the presidency of George W. Bush, the New Gilded Age with its all-encompassing social relations and punitive solutions to a wide range of social problems seemed to reach dizzying heights under an administration marked by an intensified regime of militarization, privatization, and a frenzied market fundamentalism.

While it has become fashionable to proclaim the end of history and ideology, on the one hand, and a growing public disengagement with politics, on the other, a seismic shift has taken place in the United States in the last 30 years. This shift has eviscerated the space of democratic politics as well as the language in which it is affirmed and contested. Important transformations in the nature of the state, the separation of political power from economic resources, the emergence of a market that colonizes critical agency in its own interests, and the surrender of education to the complex forces of a new electronically mediated culture reflect a new kind of sovereignty that resides in the market, outside of the constraining influence of state power. The domination of corporate sovereignty is more porous, expansive, and mobile than anything we have seen in the past and in spite of the current crisis does not appear to be in danger.[43]

I believe that we have entered into a unique theater of politics that demands a new theoretical discourse for both understanding and overcoming many of the social problems we are currently facing as a range of antidemocratic tendencies appear to be rewriting the relationship between life and politics, agency and social responsibility, and the related discourses of hope, critique, commitment, and social intervention. At stake here is the important issue of how to think about democratic politics in an age that collapses the public sphere into privatized market relations. In order to address this issue, I want to first shed light on some of the distinguishing features, inequalities, and modes of legitimation that have given rise to a New Gilded Age—which has become a code word for the sanctioning of a savage neoliberal capitalism that seeks to "destroy the very possibility of politics, freedom, and consequently, our humanity."[44]

The depoliticization of politics and the privatization of social issues are now supplemented by a bought media system that largely serves to cheerlead for neoliberal savagery, refashioning it as a combination of hip, ironic, and gross-out entertainment. This marks the emergence of a new kind of public pedagogy with extraordinary powers to influence popular consent and secure the "substitution of consumer choice for genuine political choice...albeit abetted by a lazy, corporate-run media."[45] What is so extraordinary about the New Gilded Age is how shamefully it has been celebrated in the dominant media. Stories abound about new tycoons such as William P. Foley II, the head of a major title insurance company, who is lauded for using part of his enormous wealth to buy huge tracts of land in order to preserve the natural beauty of the environment. That such land is being taken out of the public commons and turned over to a new landed gentry barely elicits a comment.[46] A complicit media also applauds stories about CEOs such as former Citigroup chairman Stanford I. Weill, who without apology states how "lucky the Carnegies and the Rockefellers were because they made their money before there was an income tax."[47] Even as the Gilded Age loses some of its glamor because of neoliberal policies and the economic nosedive, stories proliferate about how the rich are feeling the pinch as their incomes shrink from $8 million to $2 million a year. As reported in the *New York Times*, some are downright depressed because they are "cutting back on luxuries like $350 highlights and $10,000-an-hour jet rentals."[48] But jaded excess still has its appeal, at least when it comes to media representations and stories about spoiled rich kids who can be found in television shows such as *90210*, Hollywood movies such as *Beverly Hills Chihuahua*, and best-selling young adult novels such as *The Debs*, *Bratfest at Tiffany's*, and *Schooled*.[49] Unfortunately, the 13 million children who live in poverty and whose ranks are growing with the economic meltdown are not worth mentioning since they do little to increase ratings or generate media profits.

In spite of the current economic crisis, Gilded Age excess is now on display in all of the major media as a referent for "the good life." Getting ahead requires a hyped-up version of social Darwinism, endlessly played out in various "reality television" programs, which represent an insatiable and cutthroat scorn for the weaknesses of others and a sadistic affirmation of ruthlessness and steroidal power. Getting voted off the island or being told "You're fired!" now renders real-life despair and misfortune entertaining, even pleasurable. As Zygmunt Bauman points out, the dominant logic that emanates from the ongoing deluge of reality TV is clear and consistent:

[T]hat one is of use to other human beings only as long as she or he can be exploited to their advantage, that the waste bin, the ultimate destination of the excluded, is the natural prospect for those who no longer fit or no longer wish to be exploited in such a way, that survival is the name of the game of human togetherness and that the ultimate stake of survival is outliving the others. We are fascinated by what we see—just as Dali or De Chirico wished us to be fascinated by their canvases when they struggled to display the innermost, the hidden most contents of our subconscious fantasies and fear.[50]

In a celebrity-obsessed media, the obscenely wealthy offer up a seemingly inexhaustible spectacle of greed and decadence, gesturing all the while toward the arrogance of neoliberal power. Stories are published about $200 million yachts that have "multiplied from 4,000 a decade ago to 7,000 now."[51] Even the *New York Times* ran a story in the summer of 2007 providing not only a welcome endorsement of Gilded Age excess, but also barely contained praise for a growing class of outrageously rich chief executives, financiers, and entrepreneurs, described as "having a flair for business, successfully [breaking] through the stultifying constraints that flowed from the New Deal" and using "their successes and their philanthropy [to make] government less important than it once was."[52] As the United States heads "into the worst economic crisis in a half century,"[53] as the housing crisis forces millions of people to hand their homes over to the bankers, and health insurance companies adopt a pricing system in which patients, largely poor and sick, would pay hundreds if not thousands of dollars for prescriptions that may be necessary to save their lives, the *New York Times*, without irony, offered up a front-page story that focused on how the ultrarich keep right on spending, even as times get tough. Surely millions of people must have felt relieved to know that the rich were still buying $10 million condos in New York City and Italian-built yachts worth $35 million, cruising through the skies in their private jets, and once on the ground gladly handing over $3000 for a bottle of Remy Martin Louis XIII cognac.[54] There is no mention of corrupting inequities to distract readers in this story, and why should there be? After all, what's wrong with reporting some uplifting news in the midst of bad times? As Mike Davis and Daniel Bertrand Monk point out, evidence of this utopian spin on greed is shamelessly reproduced and largely celebrated in popular culture and the mass media without the slightest hint of political indignation or moral outrage. They write:

No one is surprised to read about millionaires spending $50,000 to clone their pet cats or a billionaire who pays $20 million for a brief vacation in space. And if a London hairdresser has clients happy to spend $1,500 for haircuts, then why shouldn't a beach house in the Hamptons sell for $90 million or Lawrence Ellison, CEO of Oracle, earn $340,000 an hour in 2001? Indeed, so much hyperbole is depleted in the coverage of the lifestyles of billionaires and celebrities that little awe remains to greet the truly extraordinary statistics, like the recent disclosure that the richest 1 percent of Americans spend as much as the poorest 60 million; or that 22 million factory jobs in the twenty major economies were sacrificed to the gods of globalization between 1995 and 2002; or that rich individuals currently shelter a staggering $11.5 trillion (ten times the annual GDP of the UK) in offshore tax havens.[55]

As Davis and Monk suggest, one key characteristic of the now devalued New Gilded Age is that wealth is being concentrated among the richest groups in the United States at an alarming rate, giving rise to a gap between the rich and the poor unlike anything replicated since the 1920s.[56] For instance, "In 2004, the richest 1 percent in the United States held over $2.5 trillion more in net worth than the entire bottom 90 percent.... In 1976, the top 1 percent of the population received 8.83 percent of national income. In 2005, they grabbed 21.93 percent."[57] Peter Dreier contextualizes these figures in an ever-expanding set of indices and network of inequality, if not class warfare. He writes:

Today, the richest one percent of Americans has 22 percent of all income and about 40 percent of all wealth. This is the biggest concentration of income and wealth since 1928. In 2005, average CEO pay was 369 times that of the average worker, compared with 131 times in 1993 and 36 times in 1976. At the pinnacle of America's economic pyramid, the nation's 400 billionaires own 1.25 trillion dollars in total net worth—the same amount as the 56 million American families at the bottom half of wealth distribution.[58]

As many Americans face the pressures of higher mortgage rates, soaring college tuition, galloping health care costs, and bleak possibilities for retirement savings, their quality of life plummets while executive compensation has gone through the roof. Ellen Simon reported that half of all "CEOs received compensation of more than $8.3 million a year, and some make much, much more."[59] For example, Yahoo! Inc.'s Terry Semel garnered a total compensation package in 2006 worth $71.7 million, and the top ten earners in various industries averaged $30 million each. Some hedge fund managers, with their special tax breaks, make even these amounts seem like small change. In the New Gilded Age, executive compensation begins to look like corporate fraud as in the cases of CEOs James Simons of Renaissance of Technologies Corporation, Kenneth Griffin of Citadel Investment Group and Sears Holding Corporation, and Edward Lampert of ESL Investments, who "collectively earned $4.4 billion last year."[60] Even failed CEOs get rewarded in the age of unrepentant greed. Robert Nardelli, the CEO of Home Depot, received a severance package of $210 million even though Home Depot's stock fell by 12 percent in 2006 and shareholders bore the brunt of a 40 percent decline in the stock's value, amounting to a loss of more than $25 billion.[61]

It gets worse. Charles O. Prince III, the former chief executive of Citigroup, left the company "with an exit package worth $68 million, a $1.7 million pension, an office and assistant, and a car and a driver" even though the company lost $64 billion in market value during his tenure.[62] In 2008, the company announced write-offs worth roughly $20 billion, while its market shares have plummeted over 60 percent in the last year.[63] And yet even as the public turns against these excesses at a time of financial turmoil and some of these CEOs are vilified in the public press, little is said in the mainstream media about the fundamentally antidemocratic tendencies and practices of corrupt corporate politics and a market system that produced them. The financial crisis must be understood as not simply being about individual greed writ large; it is more importantly about a crisis of legitimation regarding the value and credibility of those in charge of running the financial sectors driving neoliberal capitalism. As Paul Krugman points out, 'there's no longer any reason to believe that the wizards of Wall Street actually contribute anything positive to society, let alone justify those humongous paychecks.[64]

Of course, there is more at stake here than the emergence of a new class of rich tycoons, a predatory narcissism, a decadent materialism, and a neofeudal worldview in which the future can be measured only in immediate financial gains. There is also the growing threat to the planet as democracy is largely redefined in the interests of corporate values and profits. Corporate power translates into political power for the rich and further impoverishment for everyone else. Government policies are made into laws that benefit the rich through tax subsidies and legal protections, while undercutting, underfunding, and eliminating social protections aimed to help the poor, aged, and sick, including children. For example, in the wake of the widening housing and mortgage crisis in which home foreclosures reached over two million and hundreds of thousands of individuals and families risked losing not only their homes but also any viable place to live, President Bush and his supporters blocked a Democratic Party–backed bill that would have prevented as many as 600,000 home foreclosures, rescuing thousands of borrowers from becoming homeless. In this case, Bush's allegiance to corporate power was on full display not only with his decision to side with the banks, Wall Street firms, and mortgage lenders, but also in his response to criticism of his veto of the mortgage relief bill.[65] Rather than address the crisis, Bush shamelessly exploited it for his own ideological ends, playing politics with human tragedy by using the mortgage crisis relief efforts to call on "Congress to extend indefinitely his 2001 and 2003 tax cuts,"[66] which largely benefit the rich and powerful corporations.

But these are not the only programs providing relief for those in need that are at risk of being dismantled in the face of neoliberal policies that aim, to use Grover Norquist's memorable phrase, "to get the federal government down to the size where we can drown it in the bathtub."[67] Programs that actually work to benefit people in need are in grave danger, such as Social Security, Medicare, and even successful programs for millions of poor kids such as the popular State Children's Health Insurance Program (SCHIP), the legislation for which President Bush vetoed a number of times before a watered-down version of the bill was passed. One influential conservative, William Kristol, editor of the *Weekly Standard* and an op-ed writer for the *New York Times*, quipped in light of the public outrage over Bush's veto of the bill, "First of all, whenever I hear anything described as a heartless assault on our children, I tend to think it's a good idea. I am happy that the President's willing to do something bad for kids." Maybe this is just bad irony at work, but the joke reveals, as Paul Krugman points out, an underlying premise largely believed by the new market fundamentalists of the Gilded Age that "only wimps actually care about the suffering of others."[68]

This type of malignant arrogance parading as humor gains its legitimacy in a culture where there is no longer a language for democracy and no discourse for talking about social investments, protection, responsibility, community, and engaged citizenship. In part, social exclusions, widespread human suffering, and ongoing collective misfortune are now not only dismissed with a snicker and allotted to the discourse of faulty character, individual irresponsibility, or just plain laziness, but also viewed as essential for the success of the system and the few who benefit from it. Social death, disposability, and the promotion of human waste represent more than exceptional moments in an otherwise efficient neoliberal marketplace; such elements point to what is now commonplace—what is indeed central to the current Gilded Age (even in its weakened state)—its promotion of a ruthless politics in which the categories of social justice, citizenship, democracy, and the public good are either barely acknowledged or dismissed with contempt. Massive disparities in wealth and influence along with the weakening of worker protections and the destruction of the social state are now legitimated through a rewriting of history in which political power is measured by the degree to which it evades any sense of actual truth and moral responsibility. In this case neoliberalism not only makes power invisible but also erases a history of barbaric greed, unconscionable economic inequity, rapacious Robber Barons, scandal-plagued politics, resurgent monopolies, and an unapologetic racism.

A marauding market fundamentalism now rules most aspects of social life in the developed world, if not the globe, and the mutually constitutive forces of terror and corporate-enriching values are now becoming the regulative principles of everyday life. Without a hint of irony, it is apparent that "neoliberalism sees the market as the very paradigm of freedom, and democracy emerges as a synonym for capitalism, which has reemerged as the telos of history."[69] Global flows of capital now work in tandem with a deference for all things military, while democracy is invoked to function as a transparent legitimation for empire abroad under the conceit of political expediency in the "war on terror," which mimics the tawdriness and deceit of a rampant culture of corruption and secrecy at the highest levels of government.[70] As finance capital reigns supreme over American society, bolstered by "the new and peculiar power of the information revolution in its electronic forms,"[71] democratization along with the public spheres needed to sustain it becomes an unsettled and increasingly fragile, if not endangered, project. Put differently, the Gilded Age not only returned with a vengeance but, in spite of the widespread misery it has caused, still displays the great extremes of wealth and human suffering with a mocking haughtiness that suggests both an utter disregard for the widespread hardships it promotes and a gloating arrogance on the part of those elites who are benefiting from the new extremes of wealth and power.[72] How else to explain American International Group giving away $165 million in bonuses to top executives after it received a $180 billion in bailout funds from the Obama administration? Moreover, the perfect storm of arrogance and greed was on display at it gave the bonuses to the very division that was the source of AIG's collapse.

I have outlined above some of the excesses and antidemocratic tendencies of the New Gilded Age within a specific analysis of neoliberal rationality, its inequalities of wealth, and the economic, cultural, and political ideologies that enable its ongoing production and legitimation. In what follows, I will draw upon a number of theoretical discourses that deal with the related Foucauldian themes of governmentality and biopolitics in order to explore further how neoliberalism produces a distinct logic of disposability that increasingly constitutes those marginalized by poverty, race, and age as human waste to be rendered either invisible, redundant, or perishable. In doing so, I will argue that a theory of biopolitics offers new possibilities both for critically engaging what Nick Couldry calls the "theatre of cruelty"[73] that defines a free-market neoliberalism, especially in terms of its effects on youth outlined in the previous chapters, and for critically assessing the shifts in its underlying regime of politics, including most importantly how it can be challenged and overcome in the interests of a substantive and flourishing democracy.

## Neoliberalism as a Biopolitics of Disposability

The mutually determining forces of ever-deepening inequality and an emerging repressive state apparatus have become the defining features of neoliberalism at the beginning of the new millennium. Wealth is now redistributed upward to produce record-high levels of inequality, and corporate power is simultaneously consolidated at a speed that threatens to erase the most critical gains made over the last 50 years to curb the antidemocratic power of corporations. Draconian policies aimed at hollowing out the social state are now matched by an increase in repressive legislation to curb the unrest that might explode among those populations falling into the despair and suffering unleashed by a "savage, fanatical capitalism" that now constitutes the neoliberal war against the public good, the welfare state, and "social citizenship."[74] Privatization, commodification, corporate mergers, and asset stripping go hand in hand with the curbing of civil liberties, the increasing criminalization of social problems, and the fashioning of the prison as the preeminent space of racial containment (one in nine black males between the ages of 20 and 34 are incarcerated).[75] The alleged morality of market freedom is now secured through the ongoing immorality of a militarized state that embraces torture, war, and violence as legitimate functions of political sovereignty and the ordering of daily life. As the rich get richer, corporations become more powerful, and the reach of the punishing state extends itself further, the forces and public spheres that once provided a modicum of protection for workers, the poor, the sick, the aged, and the young are undermined, leaving large numbers of people impoverished and with little hope for the future.

David Harvey refers to this primary feature of neoliberalism as "accumulation by dispossession,"[76] which encompasses the privatization and commodification of public assets, deregulation of the financial sector, and the use of the state to direct the flow of wealth upward through, among other practices, tax policies that favor the rich and cut back the social wage. As Harvey points out, "All of these processes amount to the transfer of assets from the public and popular realms to the private and class privileged domains"[77] and to the overwhelming of political institutions by powerful corporations that keep them in check. Zygmunt Bauman goes further and argues that not only does capitalism draw its lifeblood from the relentless process of asset stripping, but it produces "the acute crisis of the 'human waste' disposal industry, as each new outpost conquered by capitalist markets adds new thousands or millions to the mass of men and women already deprived of their lands, workshops, and communal safety nets."[78] The upshot of such policies is that larger segments of the population are now struggling under the burden of massive debts, bankruptcy, unemployment, lack of adequate health care, and a brooding sense of hopelessness. What is unique about this type of neoliberal market fundamentalism is not merely the antidemocratic notion that the market should be the guide for all human actions, but also the sheer hatred for any form of sovereignty in which the government could promote the general welfare.

As Thom Hartmann points out, governance under the regime of neoliberalism has given way to punishment as one of the central features of politics. He describes the policies endorsed by neoliberals as follows: "Government should punish, they agree, but it should never nurture, protect, or defend individuals. Nurturing and protecting, they suggest, is the more appropriate role of religious institutions, private charities, families, and—perhaps most important—corporations. Let the corporations handle your old-age pension. Let the corporations decide how much protection we and our environment need from their toxins. Let the corporations decide what we're paid. Let the corporations decide what doctor we can see, when, and for what purpose."[79] But the punishing state does more than substitute charity and private aid for government-backed social provisions, while it criminalizes a range of existing social problems. It also cultivates a culture of fear and suspicion toward all those others—immigrants, refugees, Muslims, youth, minorities of class and color, the unemployed, the disabled, and the elderly—who in the absence of dense social networks and social supports fall prey to unprecedented levels of displaced resentment from the media, public scorn for their vulnerability, and increased criminalization because they are considered both dangerous and unfit for integration into American society.

Coupled with this rewriting of the obligations of sovereign state power and the transfer of sovereignty to the market is a widely endorsed assumption that regardless of the suffering, misery, and problems faced by human beings, they are not only responsible for their fate but reliant ultimately on themselves for survival. There is more going on here than the vengeful return of an older colonial fantasy that regarded the natives as less than human, or the figure of the disposable worker as a prototypical by-product of the capitalist order—though the histories of racist and class-based exclusion inform the withdrawal of moral and ethical concerns from these populations.[80] What we are currently witnessing is the unleashing of a powerfully regressive symbolic and corporeal violence against all those individuals and groups who have been "othered" because their very presence undermines the engines of wealth and inequality that drive the neoliberal dreams of consumption, power, and profitability.

The complex nature of sovereignty today signals the emergence of a fundamentally new mode of politics in which state power not only takes on a different register but in many ways has been modified by the sovereignty of the market. While the state still has the power of the law to reduce individuals to impoverishment and to strip them of civic rights, due process, and civil liberties, neoliberalism increasingly wields its own form of sovereignty through the invisible hand of the market, which now has the power to produce new configurations of control, regulate social health, and alter human life in unforeseen and profound ways. Zygmunt Bauman's analysis of how market sovereignty differs from traditional modes of state sovereignty is worth citing in full.

This strange sovereign [the market] has neither legislative nor executive agencies, not to mention courts of law—which are rightly viewed as the indispensable paraphernalia of the bona fide sovereigns explored and described in political science textbooks. In consequence, the market is, so to speak, more sovereign than the much advertised and eagerly self-advertising political sovereigns, since in addition to returning the verdicts of exclusion, the market allows for no appeals procedure. Its sentences are as firm and irrevocable as they are informal, tacit, and seldom if ever spelled out in writing. Exemption by the organs of a sovereign state can be objected to and protested against, and so stands a chance of being annulled—but not eviction by the sovereign market, because no presiding judge is named here, no receptionist is in sight to accept appeal papers, while no address has been given to which they could be mailed.[81]

Traditional political sovereignty recognized its dependency on the people it governed and to whom it remained accountable. But no one today votes for which corporations have the right to dominate the media and filter the information made available to the public; there is no electoral process that determines how private companies grant or deny people access to adequate health care and other social services. The reign of the market in a neoliberal economy is not restricted to a limited term of appointment, despite the market's unprecedented sovereignty over the lives of citizens in democratic countries—sovereignty essentially defined as the "power and capacity to dictate who may live and who may die."[82] This shift to market sovereignty, values, and power points to the importance of *biopolitics* as an attempt to think through not only how politics uses power to mediate the convergence of life and death, but also how sovereign power proliferates those conditions in which individuals marginalized by race, class, and gender configurations are "stripped of political significance and exposed to murderous violence."[83]

The notion that biopolitics marks a specific moment in the development of political modernity has been expounded in great detail by Michel Foucault.[84] Foucault argues that since the seventeenth and eighteenth centuries, with emerging concerns for the health, habitation, welfare, and living conditions of populations, the economy of power is no longer primarily about the threat of taking life, or exercising a mode of sovereign power "mainly as a means of deduction—the seizing of things, time, bodies, and ultimately the seizing of life itself."[85] For Foucault, biopolitics points to new relations of sovereignty and power that are more capacious, concerned not only with the body as an object of disciplinary techniques that render it "both useful and docile" but also with a body that needs to be "regularized,"[86] subject to corrective mechanisms and immaterial means of production that exert "a positive influence on life, endeavour[ing] to administer, optimize, and multiply it."[87] For Foucault, power is no longer exclusively embodied in the state or its formal repressive apparatuses and legal regulations.[88] Instead, power also circulates outside of the realm of the state and the constraints of a juridico-discursive concept—through a wide variety of political technologies and modes of subjectification, produced through what Foucault calls governmentality or the pedagogical "tactics . . . which make possible the continual definition and redefinition of what is within the competence of the state and what is not, the public versus the private, and so on."[89] In this instance, biopolitics does not collapse into sovereign power, just as matters of consent and persuasion cannot be reduced to the disciplining of the body. As the boundary between politics and life becomes blurred, human beings and the social forms and living processes through which they live, speak, act, and relate to each other move to the center of politics, just as the latter processes and relationships become the center of new political struggles. Biopolitics thus marks a shift in the workings of both sovereignty and power as made clear by Foucault, for whom biopolitics replaces the power to dispense fear and death "with that of a power to foster life— or disallow it to the point of death. . . . [Biopolitics] is no longer a matter of bringing death into play in the field of sovereignty, but of distributing the living in the domain of value and utility. Its task is to take charge of life that needs a continuous regulatory and corrective mechanism."[90] As Foucault insists, the logic of biopolitics is largely productive, though it exercises what he calls a death function when the state "is obliged to use race, the elimination of races, and the purification of the race, to exercise its sovereign power."[91]

Neoliberalism as a mode of biopolitics not only expands the sites, range, and dynamics of power relations but also points to new modes of subjectification in which various technologies connecting the self and diverse modes of domination,[92] far removed from the central power of the state, play a primary role in producing forms of consent, shaping conduct, and constituting "people in such ways that they can be governed."[93] According to Judith Butler, as a mode of governmentality, biopolitics

is broadly understood as a mode of power concerned with the maintenance and control of bodies and persons, the production and regulation of persons and populations, and the circulation of goods insofar as they maintain and restrict the life of the population.... Marked by a diffuse set of strategies and tactics, governmentality gains its meaning and purpose from no single source, no unified sovereign subject. Rather, the tactics characteristic of governmentality operate diffusely, to dispose and order populations, and to produce and reproduce subjects, their practices and belief, in relation to specific policy aims. Foucault maintained, boldly, that "the problems of governmentality and the techniques of government have become the only political issues, the only real space for political struggle and contestation."[94]

Foucault believed that the connection between life and politics coincided with the beginning of modernity, as governance became associated with but not limited to state power, and concerned more with ordering, regulating, and producing life.

Unlike Foucault, Giorgio Agamben argues that biopolitics is the founding moment of politics and dates back to the birth of sovereignty itself, while at the same time acknowledging that biopolitics "constitutes the decisive event of modernity and signals a radical transformation of the political-philosophical categories of classical thought."[95] According to Agamben, biopolitics in the current historical moment exhibits a more forceful and dangerous register of how power seizes life, targeting it as something to strategically order, control, and possibly discard. In this view, biopolitics is more ominous than Foucault suggests, taking on a narrow and menacing guise in the new millennium.[96] The secret foundation of sovereignty, the state of exception and its logic of exclusion and reduction of human beings to "bare life," has moved from the margin to the center of political life. According to Agamben, state power as a mode of biopolitics is irreparably tied to the forces of death, abandonment, and the production of "bare life,"[97] whose ultimate incarnation is the Holocaust with its ominous specter of the concentration camp. In this formulation, the Nazi death camps become the primary exemplar of control, the new space of contemporary politics in which individuals are no longer viewed as residents or citizens but are now seen as inmates, stripped of everything, including their right to live.[98] The camp now becomes "the hidden matrix of politics,"[99] understood less as a historical fact than as a prototype for those spaces that produce "bare life." As Agamben puts it, "*The camp is the space that is opened when the state of exception begins to become the rule.*"[100]

The uniting of sovereign power and bare life, the reduction of the individual to *homo sacer*—the sacred man who under certain states of exception "may be killed and yet not sacrificed"—no longer represents the far end of political life.[101] For Agamben, "Today it is not the city but rather the camp that is the fundamental bio-political paradigm of the West."[102] In this updated version of the ancient category of *homo sacer*, it is the human who stands beyond the confines of both human and divine law—"a human who can be killed without fear of punishment."[103] As modern states increasingly suspend their democratic structures, laws, and principles, the very nature of governance changes as "the rule of law is routinely displaced by the state of exception, or emergency, and people are increasingly subject to extra-judicial state violence."[104] The life unfit for life, unworthy of being lived, is no longer marginal to sovereign power but is now fundamental to its form of governance. As the camp has become "the nomos of the modern,"[105] state violence and totalitarian power, which in the past either were generally short-lived or existed on the fringes of politics and history, have now become the rule, as life is more ruthlessly regulated and placed in the hands of military and state power. This is not to suggest as some critics argue that Agamben equates liberal democracies with totalitarian states. Instead, as Thomas Lemke explains, Agamben "does not mean to reduce or negate those profound differences, but instead tries to elucidate the common ground for these very different forms of government: the production of bare life, [asking] in what sense 'bare life' is an essential part of our contemporary political rationality."[106] In the current historical moment, as Catherine Mills points out, "all subjects are at least potentially if not actually abandoned by the law and exposed to violence as a constitutive condition of political existence."[107] Agamben's claim that "biopolitics has passed beyond a new threshold—in modern democracies it is possible to state in public what the Nazi biopoliticians did not dare say"[108] certainly rang true in the United States in recent years when it seemed war had become the highest national ideal, the CIA created its own prisons called "black sites," the government kidnapped people and sent them to authoritarian countries to be tortured, American citizens were imprisoned offshore in Navy vessels without the right to legal counsel, and an imperial presidency violated international law at will while undermining constitutional law at home.

Agamben's all-important view of a state in which lawlessness becomes fundamental to the very definition of law and its expanding exercise of sovereign violence, in which all "subjects are potentially *homo sacers* [who] are at least potentially, if not actually, abandoned by the law and exposed to violence as a constitutive condition of political existence,"[109] provides both a needed sense of urgency in the face of intensified authoritarian tendencies in the United States and a rethinking of the very nature of democratic values, if not politics itself. Moreover, Agamben reminds us of the reality of state power and state terrorism at a time when it has become fashionable to suggest that the nation-state has lost much of its power under the sway of global capitalism. Agamben provides a theoretical service in making it clear that the state has not lost its power; it has simply reconfigured it, assuming all of the properties of a carceral state. While Agamben's notion of biopolitics places little emphasis on the productive nature of power or, for that matter, on power that does not originate with the state, it rightly reminds us of "how life shorn of civic and political rights" has increasingly become a primary preoccupation of modern sovereignty.[110] While the concept of "bare life" is not without serious theoretical limitations, it is partially convincing in that "the central political significance of the camp is more plausible than many of his critics admit."[111] Moreover, in spite of its theoretical problems, it both gestures toward and offers a theoretical language necessary to critically analyze the emerging logic of disposability crucial to neoliberal modes of governmentality, policies, and social relations. In fact, as the logic of the market fosters a narrow sense of responsibility, agency, and public values, it reinforces a politics of disposability in which diverse individuals and populations are not only considered redundant and disposable but barely acknowledged to be human beings.

Agamben's notion of "bare life" is recognizable in the horrific images that followed Hurricane Katrina,[112] the response to which exposed a politics of disposability that revealed the leadership of the United States to be capable of the barbarism its colonial rhetoric has often attributed to "less developed" countries in order to delegitimize their governments. During and after Katrina, TV cameras provided countless images of hundreds of thousands of poor people, mostly blacks, some Latinos, many elderly, and a few white people, stranded on rooftops, or isolated on patches of dry highway without any food, water, or any place to wash, urinate, or find relief from the scorching sun. Newspapers printed shocking stories about dead people, mostly poor African Americans, left uncollected in the streets, on porches, in hospitals, nursing homes, electric wheelchairs, and collapsed houses, prompting some people to claim that New Orleans resembled a "Third World Refugee Camp."[113] While the dominant media reduced the Katrina debacle to government incompetence, the real agenda responsible for Katrina reveals a political rationality that is closer to Agamben's metaphor of "bare life." In this case, the aftermath of Katrina revealed the emergence of a new kind of politics, one in which entire populations are considered expendable, an unnecessary burden on state coffers, and consigned to fend for themselves. At the same time, Katrina revealed what Angela Davis insists "are very clear signs of . . . fascist policies and practices," which not only construct an imaginary social environment for all of those populations rendered disposable but also exemplify a site and space "where democracy has lost its claims."[114]

The biopolitics of neoliberalism as an instance of "bare life" is not only coming more and more to the foreground but is also restructuring the terrain of everyday life for vast numbers of people. As an older politics associated with the social state and the "social contract" (however damaged and racially discriminating)[115] gives way to an impoverished vocabulary that celebrates private financial gain over human lives, public goods, and broad democratic values, the hidden inner workings[116] of "bare life" become less of a metaphor than a reality for millions of people whose suffering and misery embodies a shift on the part of the state's response from benign neglect to malign neglect. Beyond the very visible example of Katrina, there is a host of less visible instances affecting those dehumanized by a politics of disposability. The logic of disposability as an instance of "bare life" was visible in the Bush administration's indifference to the growing HIV crisis among young black women who "represent the highest percentage (56 percent) of all AIDS cases reported among women, and an increasing proportion of new cases (60 percent)."[117] In addition to the rhetoric of color blindness and self-help that assists in camouflaging the racist underpinnings of much of contemporary society, the HIV epidemic spreads but gets almost no attention from "leaders in public health, politics, or religion."[118] It remains to be seen if the Obama administration will address this problem within the next few years.

The politics of "bare life" also informs the fury of the new nativism in the United States at the dawn of the twenty-first century. Stoked by media panics and the hysterical populist rhetoric of politicians, racist commitments easily translate into policies targeting poor youth of color and immigrant men, women, and children with deportation, incarceration, and state-backed violence. Extending the logic of disposability to those defined as "other" through the discourse of nativism, citizen border patrols and "migrant hunters"[119] urge the government to issue a state of emergency to stop the flow of immigrants across the United States' southern border. Leading public intellectuals inhabit the same theoretical discourse as right-wing vigilante groups. For example, internationally known Harvard University faculty member Samuel P. Huntington unapologetically argues in *Who Are We? The Challenge to America's Identity* that Western civilization, as it is said to be represented in the United States, is threatened by the growing presence of Hispanic Americans, especially Mexican Americans, shamelessly described as the "brown menace."[120] The *New York Times* claims that "toughness" is the new watchword in immigration policy, which translates not only into a boom in immigration detention but also in some cases into death to immigrants denied access to essential medicines and health care.[121] This nativist racism has become even more shrill in light of the H1N1 pandemic. The new biopolitics of disposability is further evident in the fact that for many black youth, the war on drugs and crime signals the emergence of "the prison as the preeminent U.S. racial space."[122] Biopolitics combines with biocapital in one of its most ruthless expressions as the carceral state increasingly runs for-profit prisons and uses inmates in prison jobs that provide profits for private contractors while exposing the prisoners to "a toxic cocktail of hazardous chemicals."[123] The logic of disposability as an instance of "bare life" also gains expression in the slavery-like conditions many guest workers endure in the United States. Routinely cheated out of wages, held captive by employers who seize their documents, and often forced to live in squalid conditions without medical benefits, such workers exist in a state that Congressman Charles Rangel characterizes as "the closest thing I've ever seen to slavery," a comment amply supported by the Southern Poverty Law Center report *Close to Slavery: Guestworker Programs in the United States.*[124] The growing armies of the "living dead" also include the 750,000 who are homeless in America on any given night,[125] along with the swelling ranks of the working poor and, the 6 million unemployed since the recession began in December 2007, and the additional nine million who have lost employer-sponsored insurance and are unable to get even minimum health care. Needless to say, the logic of disposability as an instance of "bare life" is clearly visible in all of these examples.

As important as Agamben's work is in locating matters of "life and death within our ways of thinking about and imagining politics,"[126] it needs to be analyzed, supplemented, and rethought in relation to a number of theoretical shortcomings. Since there is a vast literature available on this question, I will only briefly mention some of the more crucial concerns that need to be addressed in any viable challenge to neoliberalism and its biopolitics of disposability. First, Agamben's notion of sovereignty is too state-centered and overly associated with state power. He offers no analysis of how decisions about life and death have now been appropriated by the sovereignty of the market. The sites of sovereignty and power are multiple, suggesting a restructuring of both the state since the Nazi era and the modalities of power itself. As Thomas Lemke puts it, "Contemporary biopolitics is essentially political economy of life that is neither reducible to state agencies nor to the form of law. Agamben's concept of biopolitics remains inside the ban of sovereignty, it is blind to all the mechanisms operating beneath or beyond the law."[127] Under the reign of neoliberalism, biopolitical mechanisms extend far beyond those deprived of legal rights, encompassing a much broader range of individuals and groups subject to the "social processes of exclusion—even if they may be formally enjoying full political rights: the 'useless,' the 'unnecessary,' or the 'redundant.'"[128] Second, while Agamben wants to put into view the army of disposable beings stripped of their rights and provide a more complex, if not urgent, analytic for "the ways and forms of [how] a new politics must be thought," his notion of politics can only imagine sovereign power as a mode of domination and neglects to consider what Ernesto Laclau describes as "the system of possibilities that such a structure opens."[129] There is no hint of counterstruggles in this discourse, no examples of individual and social resistance, no suggestion of moral and political indignation leading to oppositional social movements. Sovereignty appears in Agamben's framework to collapse into unadulterated domination, eliminating rather than rethinking a space for politics. Third, there is a homogenizing logic to Agamben's "grand allegories of exclusion, crisis, and apocalypse"[130] that fails to capture how "bare life is implicated in the gendered, sexist, colonial, and racist configurations of biopolitics."[131] The result is that while Agamben provides a theoretical service in examining the emergence of new forms of domination, he overlooks how modes of domination such as sexual violation, modern-day slavery, and racism not only offer a complex genealogy of domination but might also provide a rich historical and contemporary legacy of individual challenges and collective resistance to the "bare life" politics of sovereign exclusion. Contemporary power relations, the politics they produce, and the modes of resistance they engender are far more complex than Agamben indicates. The march toward a totalitarian society is not inevitable. This overdetermination in Agamben's work does more than undercut a politics of possibility: it unwittingly feeds into an already pervasive form of political cynicism and nihilism. Ernesto Laclau is worth repeating at length on this issue:

The myth of a fully reconciled society is what governs the (non-)political discourse of Agamben. And it is also what allows him to dismiss all political options in our societies and to unify them in the concentration camp as their secret destiny. Instead of deconstructing the logic of political institutions, showing areas in which forms of struggle and resistance are possible, he closes them beforehand through an essentialist unification. Political nihilism is his ultimate message.[132]

At the dawn of the new millennium, it is commonplace for references to the common good, public trust, and public service to be either stigmatized or sneered at by people who sing the praises of neoliberalism and its dream of turning "the global economy . . . into a planetary casino."[133] Against this dystopian condition, the American political philosopher Sheldon Wolin has argued that because of the increasing power of corporations and the emergence of a lawless state (given immense power during the administration of George W. Bush), American democracy is not only in crisis but is also characterized by a sense of powerlessness and experience of loss that must be challenged. Wolin claims that this sense of loss is related "to power and powerlessness and hence has a claim upon theory."[134] In making a claim upon theory, loss aligns itself with the urgency of a crisis, a crisis that demands a new theoretical discourse while at the same time requiring a politics that involves contemplation, that is, a politics in which modes of critical inquiry brush up against the more urgent crisis that threatens to shut down even the possibility of critique. For Wolin, the dialectic of crisis and politics points to three fundamental concerns that need to be addressed as part of a broader democratic struggle. First, politics is now marked by pathological conditions in which issues of death are overtaking concerns with life. Second, it is no longer possible to assume that democracy is tenable within a political system that daily inflicts massive suffering and injustices on weak minorities and those individuals and groups who exist outside of the privileges of neoliberal values, that is, those individuals or groups who exist in what Achille Mbembe calls "death-worlds, new and unique forms of social existence in which vast populations are subjected to conditions of life conferring upon them the status of the living dead."[135] Third, theory in some academic quarters now seems to care more about matters of contemplation and judgment in search of distance rather than a politics of crisis driven by an acute sense of justice, urgency, and intervention. Theory in this instance distances itself from politics, neutered by a form of self-sabotage in which ideas are removed from the messy realm of politics, power, and intervention. According to Wolin, "Even though [theory] makes references to real-world controversies, its engagement is with the conditions, or the politics, of the theoretical that it seeks to settle rather than with the political that is being contested over who gets what and who gets included. It is postpolitical."[136]

Wolin's emphasis on reclaiming modes of theorizing that focus on that which is positive and life-affirming in the current march toward a society that increasingly treats individuals and groups who are dispossessed, excluded, and expendable as human waste holds great promise for addressing both the biopolitics of neoliberalism and the related crisis of democracy. The emerging discourse of biopolitics, especially in Foucault, Agamben, and more recently in Michael Hardt and Antonio Negri, is invaluable in taking up Wolin's challenge.[137] While Agamben and others speak powerfully and thoughtfully to a biopolitics in which the thought of living a decent life is now supplanted by the task of either surviving or outwitting death, Foucault offers a more compelling notion of power and biopolitics, one that inscribes power in multiple sites and invests it with the possibility of doing far more politically than simply producing bare life. Foucault's theory of power offers an alternative to Agamben's notion of power as centralized and repressive, emphasizing instead how the productive side of power emerges along with the historical and political novelty of diverse modes of resistance and struggle.[138] Foucault insists that the logic of biopolitics "exerts a positive influence on life, endeavours to administer, optimize, and multiply it."[139] Yet while Foucault understands the singular importance of biopolitics to foster life as its object and objective, making, producing, and expanding the possibility of what it means to live as its central and primary function, he also argues that biopolitics does not remove itself from "introducing a break into the domain of life that is under power's control: the break between what must live and what must die."[140]

Foucault is acutely aware that the impoverishment of the social order is fed by a society that neither questions itself nor can imagine any alternative to itself, and that such a rationality not only yields a partial apprehension of how power works but also feeds the growing ineptitude, if not irrelevance, of (in)organic and traditional intellectuals, whose cynicism often translates into complicity with the forms of power they condemn. Moreover, Foucault was deeply committed to analyzing how technologies of power produce particular rationalities, modes of identification, conduct, and orders of consent, and how they were mediated and integrated through "techniques of the self and structures of coercion and domination."[141] Foucault's notion of "governmentality"[142] suggests that, as Lemke argues, "it is important to see not only whether neoliberal rationality is an adequate representation of society but also how it functions as a 'politics of truth,' producing new forms of knowledge, inventing different notions and concepts that contribute to the 'government' of new domains of regulation and intervention."[143] While Foucault does not use the term pedagogy, his notion of governmentality is extremely suggestive regarding the importance of making pedagogy crucial to any notion of politics.

Michael Hardt and Antonio Negri build upon Foucault's theoretical insights by emphasizing biopolitics as a productive "form of power that regulates social life from its interior."[144] Hardt and Negri argue that biopolitics not only touches all aspects of social life but is the primary political and cultural force through which the creation and reproduction of new subjectivities take place, while registering culture, society, and politics as a terrain of multiple and diverse struggles waged by numerous groups. In this perspective, biopolitics is mediated through the world of ideas, knowledge, new modes of communication, and a proliferating multitude of diverse social relations. Hardt and Negri argue that this ample notion of biopolitics registers a global world in which production is not merely economic but social—"the production of communications, relationships, and forms of life" that allows diverse individuals and groups "to manage to communicate and act in common while remaining internally different," yet sharing a common currency in the desire for democracy.[145] According to Hardt and Negri, "Who we are, how we view the world, how we interact with each other are all created through this social, biopolitical production."[146] And it is precisely within this transformed biopolitical sphere that they believe new and diverse social subjects sharing a common project of resistance and democracy can emerge on a global scale. For my purposes, the importance of Agamben's, Foucault's, and

Hardt and Negri's work on biopolitics, in spite of their distinct theoretical differences, is that they move matters of culture, especially those
aimed at "the production of information, communication, [and] social
relations ... to the center of politics itself."[147] Though they may share
little else, each of these theorists recognizes that democracy is in danger. Whereas Agamben emphasizes a dystopian biopolitics attentive
to the intensification of a widespread culture of death, Foucault and
Hardt and Negri offer us important theoretical tools for addressing
the productive and cultural/pedagogical nature of biopolitics.

While the concept of pedagogy is implicit in the approaches
to biopolitics discussed above, it is underdeveloped theoretically,
particularly around matters of agency, critical consciousness, and
resistance. Building upon this absence productively suggests making pedagogy more central to any oppositional notion of biopolitics,
governmentality, and struggle. In the final section of this chapter,
I selectively appropriate elements of the work of these four theorists in order to develop an analysis of contemporary biopolitical
investments, which in my view offers the best means for challenging the insidious complexity of neoliberalism's logic of disposability.
In addition, I want to draw upon the work of a number of theorists
who make critical pedagogy—the articulation of critical knowledge to
experience—central to any viable notion of politics and critical agency.

## Conclusion: Toward a Global Democratic Public Sphere

Any attempt to address the current biopolitics of neoliberalism and
disposability must begin by decoupling what has become a powerful hegemonic element in neoliberal rationality—the presupposition
that the market is synonymous with democracy and the final stage
in "the telos of history."[148] Against this ideological subterfuge, it is
crucial for intellectuals and others not only to reveal neoliberalism as
a historical and social construction but also to make clear the various ways in which its regime of truth and power is being resisted
by other countries, particularly as "its magic seems to have faded
in the laboratories of the south, especially in Latin America, where
once Argentina, Bolivia, Brazil, and Ecuador were crowded together
as its poster children."[149] Equally important is the necessity to make
visible and critically analyze the matrix of ideological and economic
mechanisms at work under neoliberalism and how it is producing the
growing inequality of wealth and power throughout the globe, as well
as the current worldwide economic crisis.[150]

And yet revealing the specific material relations underlying the power, institutions, and rationality at work in the biopolitics of neoliberalism, while important, is not enough. What must also be addressed in resisting the biopolitics of neoliberalism is its concerted assault on the very existence of politics and democracy, and the educational conditions that make them possible. Central to such a challenge is the necessity to address how neoliberalism as a pedagogical practice and a public pedagogy operating in diverse sites has succeeded in reproducing in the social order a kind of thoughtlessness—a social amnesia of sorts—that makes it possible for people to look away as an increasing number of individuals and groups are made disposable, relegated to new zones of exclusion marked by the presupposition that life is cheap, if not irrelevant, next to the needs of the marketplace and biocapital. Of course, there is more at stake here than providing an ideological analysis of neoliberal economics, politics, and hegemony; there is also, as Zygmunt Bauman has pointed out, a need to situate the biopolitics of neoliberalism within a growing economy of individuation and privatization, the current collapse of the social state, the transfer of power to larger global political forces, the death of long-term projects that embrace a democratic future, and a dissolution of all democratic social forms.[151] Under the reign of neoliberalism and its rabid market fundamentalism, society is no longer protected by the state. As neoliberalism reproduces with deadly results the multileveled economies of wealth and power, it also decouples economics from public life and morality from market forces, and in doing so creates with little opposition endless numbers of disposable populations who are stripped of their most basic rights and relegated to the axis of irrelevance. As Bauman points out, against the most basic principles of a viable democracy, neoliberalism produces disposable populations now considered not only "untouchables, but unthinkables." He writes:

In the habitual terms in which human identities are narrated, they are ineffable. They are Jacques Derrida's "undecidables" made flesh. Among people like us, praised by others and priding ourselves on arts of reflection and self-reflection, they are not only untouchables, but unthinkables. In a world filled to the brim with imagined communities, they are the unimaginables. And it is by refusing them the right to be imagined that the others, assembled in genuine hoping to become genuine communities, seek credibility for their own labours of imagination.[152]

This logic of disposability is about more than the extreme examples portrayed by the inhabitants of Agamben's camp. The biopolitics of disposability both includes and reaches beyond the shocking image of the overcrowded refugee camps and the new American Gulag that includes the massive incarceration of mostly people of color, special prisons for immigrants, torture sites such as Abu Ghraib, and the now infamous Camp Delta at Guantanamo Bay, Cuba. Disposable populations now include the 60 million people in the United States living one notch above the poverty line, the growing number of families living on bare government subsistence, the 46 million Americans without health insurance, the over 2 million persons incarcerated in prisons, the young people laboring under enormous debt and rightly sensing that the American dream is on life support, the workers who are one paycheck away from joining the ranks of the disposable and permanently excluded, and the elderly whose fixed incomes and pensions are in danger of disappearing.[153] On a global level, the archetypes of otherness and disposability can be found in "disease-ridden Africa," the Orientalist paradigm that now defines the Arab world, those geopolitical spaces that house the growing refugee camps in Europe, Africa, the Middle East, and North America, and those countries from Iraq to Argentina that have suffered under neoliberal economic policies in which matters of structural adjustment are synonymous with the dictates of what Naomi Klein calls "disaster capitalism."[154] The camp increasingly becomes the exemplary institution of global neoliberal capital—succinctly defined by Zygmunt Bauman as "garrisons of extraterritoriality," functioning largely as "dumping grounds for the indisposed of and as yet unrecycled waste of the global frontier-land."[155]

A biopolitics that struggles in the name of democratic education and politics becomes impossible unless individual and political rights are protected and enabled by social rights. As I emphasize throughout this book, this means in part that collective opposition to the punishing state and the sovereignty of the market has to be waged in the name of a democracy that takes up the struggle for a social state that not only provides social protections and collectively endorsed insurance but also redistributes wealth and income so as to eliminate the inequalities that fuel and reproduce the power of neoliberalism and its war on the welfare state, its promotion of an expanded military, its contracting out of major public services, and its call for a law-and-order state of (in)security.

Biopolitics as a concept in this struggle is essential because it makes visible a neoliberal regime in which politics makes life itself a site of radical unequal struggle. The power of global capital produces a politics of disposability in which exclusion and death become the only mediators of the present for an increasing number of individuals and groups. If the exclusion of vast numbers of people marginalized by race, class, age, and gender was once the secret of modernity, neoliberalism has amplified its power to exclude large numbers of diverse groups from a meaningful social existence, while making the logic of disposability central to its definition of politics, and, as I have also argued, its modes of entertainment. There is something more distinctive about contemporary neoliberal biopolitics than an obsession with necropolitics, where the state of exception becomes routine, a war against terrorism mimics that which it opposes, and death-dealing modes of inequality are strengthened. Neoliberalism's politics of disposability is not merely maintained through disciplinary and regulatory powers but also works primarily as a form of seduction, a pedagogy in which matters of subjectification, desire, and identities are central to neoliberalism's mode of governing.

Pedagogy, understood as a moral and political practice, functions as a form of cultural politics and governmentality that takes place in a variety of sites outside of schools. In this instance, pedagogy anchors governmentality in a "domain of cognition" functioning largely as "a grid of insistent calculation, experimentation, and evaluation concerned with the conduct of conduct."[156] Beyond securing the "domains of cognition" that shape common sense, neoliberalism also produces a pedagogy of fantasy and desire, creating a kind of "emotional habitus" through the ever-present landscapes of entertainment.[157] This pedagogical apparatus and mode of seduction proffered in the name of entertainment initiates spectators to watch an unfolding "theatre of cruelty"—to laugh at exclusion and humiliation—rather than to be moved to challenge it. And this intersection of pedagogy and politics is one area in which neoliberalism can and must be challenged.

Opposing neoliberalism, in part, suggests exposing the myths and conditions that sustain the shape of late modern politics as an economic, social, and pedagogical project. This means addressing neoliberalism as both a mode of rationality and an unprecedented intersection of governmentality and sovereignty that shapes every aspect of life. Engaging neoliberalism as a mode of governmentality that produces consent for its practices in a variety of sites requires that educators and others develop modes of pedagogical and political interventions that situate human beings as critically engaged social agents capable of addressing the meaning, character, fate, and crisis of democracy. Against a biopolitics of neoliberalism and its antidemocratic tendencies, educators, artists, intellectuals, and others might consider selectively reclaiming John Dewey's notion of democracy as an ethical ideal and engaged practice informed by an active public open to debate, dialogue, and deliberation.[158] Dewey rejected any attempt to equate democracy and freedom with a market society, and he denounced procedural definitions of democracy that he felt reduced it to the periodic rituals of elections, conceding meaningful actions to formal political institutions. According to Dewey, democracy was "a way of life" that demanded work as well as a special kind of investment, desire, and willingness to fight those antidemocratic forces that produced what he called the "eclipse of the public."[159] Dewey believed that democracy demanded particular competencies, modes of understanding, and skills that enabled individuals both to defend certain institutions as vital public spheres and to equate public freedom with the capacity for debate and deliberation and a notion of politics that rejects any commitment to absolutes. If democracy was to survive, Dewey argued that it had to be nourished by pedagogical practices that enabled young people and others to give it the kind of active and constant attention that makes it an ongoing, never-ending process of replenishment and struggle.

Hannah Arendt built upon Dewey's concerns about what it means not only to rethink the meaning of substantive democracy but also to put into place those pedagogical conditions that enable people to speak from a position of critical agency and to challenge modes of authority that speak directly to them. While Arendt did not provide a theory of pedagogy, she argued passionately about connecting any viable notion of democracy with an educated public. For her, neither democracy nor the institutions that nourished it could flourish in the absence of individuals who could think critically, exercise judgment, engage in spirited debate, and create those public spaces that constitute "the very essence of political life."[160] Arendt recognized that any viable democratic politics must offer an informed and collective challenge to modes of totalitarian violence legitimated through appeals to safety, fear, and the threat of terrorism. She wrote: "Terror becomes total when it becomes independent of all opposition; it rules supreme when nobody any longer stands in its way. If lawfulness is the essence of non-tyrannical government and lawlessness is the essence of tyranny, then terror is the essence of totalitarian domination."[161]

If, as both Arendt and Dewey argued, human beings become superfluous in societies that eliminate the conditions for debate and critical engagement, it is all the more important to once again rethink the relationship between democracy and politics in an age that relegates ethics along with the social state to the dustbin of history. Arendt believed that persuasion, reflective judgment, and debate were essential to politics, while Dewey viewed democracy as an enterprise that could be kept alive only as a consequence of an ongoing struggle to preserve it. Both Arendt and Dewey, writing in the shadow of a number of twentieth-century totalitarian regimes, knew that democracy was fragile, offered no guarantees, and could be sustained only through a democratic ethos that was nourished and cultivated in a diverse number of active public spheres.[162] And it is precisely in the struggle over a democratic ethos that modes of resistance need to be mobilized that refuse and transform the narratives, values, and seductions of the neoliberal ethos. What is needed today are growing modes of global resistance, increases in humanitarian aid, escalating calls for more rights legislation, and an expanding influence of international law.[163] Such resistance cannot be mobilized simply through ideas. Also needed is the promise and reality of public spheres that in their diverse forms, sites, and content offer pedagogical and political possibilities for strengthening the social bonds of democracy, that is, new spaces from which to cultivate the capacities for critical modes of individual and social agency as well as crucial opportunities to form alliances in the collective struggle for an oppositional biopolitics that expands the scope of vision, operations of democracy, and the range of democratic institutions. In other words, what is needed is a biopolitics that serves as a rally cry for social movements willing to fight against the terrors of totalitarianism in its various fundamentalisms and guises. A genuinely global public sphere is about more than legal rights guaranteeing freedom of speech; it is also a site that demands a certain kind of citizen informed by particular forms of education, a citizen whose education provides the essential conditions for democratic public spheres to flourish. Along with Dewey and Arendt, Cornelius Castoriadis, the great philosopher of democracy, argued that if public space is not to be experienced as a private affair, but as a vibrant sphere in which people experience and learn how to participate in and shape public life, it must be shaped through an education that provides the decisive traits of courage, responsibility, and respect, all of which connect the fate of each individual to the fate of others, the planet, and global democracy.[164]

As the intersection of life and politics becomes more pronounced, a progressive biopolitics also points to a discourse of possibility where bare life and legal exception are not the norm, where the world is no longer allotted without resistance to the winners of globalization, and where individualism and consumerism no longer provide the only sense of possibility, freedom, meaning, and responsibility. Whether inscribed in the state or in the market, predatory power—in its practice "to exclude, to suspend law, to strip human existence of civic rights and social value"[165]—is never totalizing, sutured, or incapable of being resisted and contested. As Randy Martin points out, neoliberalism along with the public violence it produces does not exhaust all other notions of political expression. But he warns against a cynicism that allows a biopolitics of death to overtake a biopolitics of life:

Without discounting the scope of repression by which the state operates, or the ever more elaborate production of relative surplus populations, consolidating all politics around the figure of death is also a tremendous narrowing of the whole range of social contestations over forms of life and the shape of society... this conception of the political is, at the very least, complicit.[166]

Jean Comaroff also finds the cynicism pervading notions of biopolitics such as Agamben's concept of "bare life" risks cutting off ongoing struggles and forms of resistance before they can be understood and explored. She emphasizes the importance of retaining hope and not capitulating to the pessimism endemic to Agamben's negative biopolitics: "If, for Agamben, a fixation on biopolitics [of death] is the defining feature of modernity *tout court*, how are we to account for the struggles currently underway over the definition of life itself, over the ways that it is mediated, interpreted, abstracted, patented?"[167]

Critical thought and action often emerge under the most oppressive conditions, giving voice to a notion of hope that feeds the language of critique and produces actions often considered previously unimaginable. Needless to say, invoking hope must be connected to a version of biopolitics in which life—meaningful, purposeful, and dignified life, not simply bare life—is both affirmed and made central to the challenge of addressing the problem of disposability as global in its roots and transformation. This suggests a political pedagogy in which injustices on a local level are linked to broader global forces, and a notion of public responsibility in which matters of human waste and disposability are "condemned not because a law is broken, but because people have been hurt."[168] In a market-driven society in which disposability is now central to modes of regulation, growth, and power, the price that is being paid in human costs is so high as to potentially spell the eventual destruction of the planet itself.

The return of Gilded Age excess with its biopolitics of wealth, greed, and gross inequality reveals its link to a historical past in which the rich squander valuable resources and remove themselves from the violence, loss, pain and death visited daily on billions of other people on the planet. But the return of the Gilded Age must be viewed not merely as a referent for history repeating itself on an expanded scale but also as a reality demanding radical critique and collective struggles for democracy—now made all the more possible due to the conditions arising from the current financial crisis, which in its devastating effects has nevertheless beneficially exposed many of the ideological contradictions and abusive power relations inherent to free-market fundamentalism. Just as suffering can no longer be treated as either routine or commonsensical, the New Gilded Age and its institutional formations, values, corruptions, and greed must be rewritten in the discourse of moral outrage, economic justice, and organized resistance. Against the apocalyptic "dream-worlds" of neoliberalism, educators and others need to find new ways to rebuild those deserted public spheres—from the schools to the media to cyberspace—where it becomes possible to produce the conditions in which individual empowerment is connected to the acquisition of not only knowledge and skills but also social power.[169]

In an age marked by outsourcing, uncertainty, deregulation, privatization, and downsizing, hope is in short supply because many people have little sense of a different future or of what it means to seek justice collectively rather than individually, relying instead on their own meager resources to combat problems that far exceed individual solutions. As shared fears, insecurities, and uncertainties replace shared responsibilities, those who bear the effects of negative globalization and neoliberalism increasingly retreat into the narrowly circumscribed worlds of either consumerism or the daily routines of struggling to survive. Ignorance, indifference, and apathy provide the conditions for political inaction and the atrophy of democratic politics. As Zygmunt Bauman insists, this withdrawal from politics and the public realm does not augur well for democracy. He writes:

But democratic politics cannot survive for long in the face of citizens' passivity arising from political ignorance and indifference. Citizens' freedoms are not properties acquired once and for all; such properties are not secure once they are locked in private safes. They are planted and rooted in the sociopolitical soil and it needs to be fertilized daily and will dry out and crumble if it is not attended to day in day out by the informed actions of a knowledgeable and committed public. It is not only the *technical* skills that need to be continually refreshed, not only the *job-focused* education that needs to be lifelong. The same is required, and with still greater urgency, by education in *citizenship*.[170]

Under such circumstances, it is more crucial than ever to develop a biopolitics of resistance that echoes Theodor Adorno's argument that "the undiminished presence of suffering, fear, and menace necessitates that the thought that cannot be realized should not be discarded.... [that theory] must come to know, without any mitigation, why the world—which could be paradise here and now—can become hell itself tomorrow."[171] If Adorno is right, and I think he is, the task ahead is to fashion a more critical and redemptive notion of biopolitics, one that takes Agamben's warning seriously about social death becoming the norm rather than the exception for most of the world's population and at the same time refuses to accept, even in its damaged forms, a glittering New Gilded Age and its underbelly—a culture of fear that promotes air-tight forms of domination. We need new political and educational narratives about what is possible in terms of producing a different future, what it means to promote new modes of global responsibility, and what it takes to create sites and strategies in which resistance to neoliberal biopolitics becomes possible. And while Barack Obama's election offers a historic opportunity for change, it would be facile to assume that a change in political leadership signals the end of the New Gilded Age along with the economic, financial, and military forces that fuel it, or that a new U.S. administration could single-handedly put a halt to the relentless expansion of neoliberal rationality and politics across the globe.

At issue in this moment of potential change is to avoid the resurgence of a politics of moral purity, a hermetic identity politics, or a narrowly defined embrace of the logic of political economy. Moreover, any viable mode of collective resistance to a biopolitics of neoliberalism must refuse a vocabulary of impoverished oppositions, one "that strives to reduce expansive vocabularies of politics, social debate, and intimacy to a straightjacket of absolute oppositions: nature and abomination, truth and concealment, good and evil."[172] Against such oversimplifications, there is a growing need for modes of critique that take seriously how power and consent shape the terrain of everyday life and that also embrace a politics of possibility that engenders a counter-biopolitics of resistance, one that moves beyond the borders of the local to rewrite the global politics in which we think and live. Jean Comaroff reminds us, "As conventional politics falters in the face of ever more elusive collaborations of wealth, power, and the law, social activism has sought to exploit the incoherence of the neoliberal order against it, finding productive footholds within the *aporias* of the market system. While it has hardly forced a capitulation on the part of governments and corporations, it has won some significant concessions."[173] There is a hint of such resistance, however diffuse, in a range of biopolitical movements that include: student groups resisting the militarization and corporatization of the university; the reconsolidation of union movements all over the globe; workers and others resisting neoliberal government reforms in Greece; a growing environmental movement; a transnational feminism; and the collective struggles of AIDS activists in Brazil, South Africa, and India. In the United States, there is newfound optimism among numerous progressive groups who see in the election of Barack Obama a guarded rejection of Bush's market fundamentalism, imperial presidency, and expansionist foreign policy. All of these movements are motivated by a new sense of political and ethical urgency, while at the same time developing anew modes of politics in which the pedagogical becomes a space of possibility both within and outside of the boundaries of the nation-state. Whether they and more radical social movements will have any effect on the Obama regime remains to be seen. What is clear is that Obama has proved to be a great deal more centrist than anyone had expected, at least as revealed by both the make up of his high-ranking advisors and many of the policies he has put into play during the first few months of his administration.

Biopolitics in its most productive moment has to be cosmopolitan, transnational in its scope, global in its sense of responsibility, and educational in its mobilizing functions. As Nick Couldry has argued, part of the struggle for a vibrant democracy requires figuring out where emergent democratic publics can be created, how practices of public connections can be sustained beyond the private sphere, and how to counter the false work of neoliberal discourse with an enlarged notion of the social and a "shared space of public action" where norms of collective agency can be developed and democratic politics sustained.[174] Engaged and substantive citizenship requires a commitment to debate, dialogue, matters of justice and power, and a willingness to listen to the claims of others.

Last but not least, one of the great challenges facing global democracy is the need to recognize and further the progressive elements that seem to bear immunity against or to break down neoliberalism's theater of cruelty, which contains tendencies that "work to destroy its own protections."[175] Any politics that takes seriously a society's ethical and political obligations to the young demands more than the production of critical knowledge and a commitment to social justice. It also suggests an ongoing struggle to create the pedagogical conditions and political sites/public spheres in which alliances can be built and global movements initiated as part of a broader effort to create new modes of identification, political subjectivities, social relations of resistance, and sources of mobilization dedicated to making the world a more humane and just space for all children. Biopolitics in the interest of global democracy is a struggle over those modes of state and corporate sovereignty that control the means of life and death. Such a struggle is both critical and redemptive, and poses an urgent challenge to all young people, educators, writers, journalists, artists, and intellectuals who claim their responsibility to stand up against the ongoing ritualistic and spectacular violence waged by neoliberalism against reason, justice, freedom, and democracy itself.

# NOTES

## INTRODUCTION

1. Sharon Stephens, "Children and the Politics of Culture in 'Late Capitalism,'" in *Children and the Politics of Culture*, ed. Sharon Stephens (Princeton: Princeton University Press, 1995), p. 7.
2. Mike Davis and Daniel Bertrand Monk, "Introduction," *Evil Paradises* (New York: The New Press, 2007), p. ix.
3. Louis Uchitelle, "The Richest of the Rich, Proud of New Gilded Age," *New York Times* (July 15, 2007). Online: http://www.nytimes. com/2007/07/15/business/15gilded.html?_r=1&oref=slogin.
4. Wendy Brown, *Edgework: Critical Essays on Knowledge and Politics* (Princeton: Princeton University Press, 2005), p. 52.
5. Zygmunt Bauman, "Happiness in a Society of Individuals," *Soundings* (Winter 2008), p. 21.
6. Tony Pugh, "US Economy Leaving Record Numbers in Severe Poverty," *McClatchy Newspapers* (February 23, 2007). Online: http://www.commondreams.org/headlines07/0223-09.htm.
7. Cesar Chelala, "Rich Man, Poor Man: Hungry Children in America," *Seattle Times* (January 4, 2006). Online: http://www.commondreams. org/views06/0104-24.htm.
8. Cited in Bob Herbert, "Children in Peril," *New York Times* (April 21, 2009), p. A25.
9. Paul Krugman, "Banks Gone Wild," *New York Times* (November 23, 2007). Online: http://www.nytimes.com/2007/11/23/opinion/ 23krugman.html?_r=1&oref=slogin&pagewanted=print.
10. Peter Dreier, "Bush's Class Warfare," *Huffington Post* (December 21, 2007). Online: http://www.huffingtonpost.com/peter-dreier/bushs-class-warfare_b_77910.html.
11. Associated Press, "Jimmy Carter Slams Bush Administration," *Dallas Morning News* (May 19, 2007). Online: http://www.dallasnews. com/sharedcontent/dwsb/news/politics/national/stories/051907 dnnatcarterbush.91ad00.html.
12. Chris Hedges, *American Fascists: The Christian Right and the War on America* (New York: Free Press, 2007).
13. Susan Buck-Morss, *Thinking Past Terror: Islamism and Critical Theory on the Left* (New York: Verso, 2003), p. 29.

14. Stanley Aronowitz, *The Last Good Job in America* (Lanham, MD: Rowman and Littlefield, 2001), p. 160.
15. Arjun Appadurai, *Fear of Small Numbers* (Durham: Duke University Press, 2006), pp. 36–37.
16. Michael Hardt and Antonio Negri, *Multitude: War and Democracy in the Age of Empire* (London: Penguin, 2004), p. 341.
17. Editorial, "So Little Time, So Much Damage," *New York Times* (November 4, 2008), p. A26.
18. Robert Pear, "Rush to Enact a Safety Rule Obama Opposes," *New York Times* (November 30, 2008), p. 1.
19. Editorial, "So Little Time, So Much Damage."
20. Jesse Jackson, "Bush's Last 100 Days the Ones to Watch," *Chicago-Sun Times* (November 4, 2008). Online: http://sudhan.wordpress.com/2008/11/05/bushs-last-100-days-the-ones-to-watch/.
21. Both quotes are taken from Michael Hardt, "Books: Best Books: Best of 2008," *Artforum* (December 2008), p. 83.
22. George W. Bush, "The Surest Path Back to Prosperity," *Man with No Name* (November 2008). Online: http://www.dallasdancemusic.com/forums/awareness-politics/276597-surest-path-back-prosperity-george-w-bush.html.
23. Ibid.
24. On the Gilded Age, see Alan Trachtenberg, *The Incorporation of America: Culture and Society in the Gilded Age* (Vancouver: Douglas and McIntyre, 2007); Michael McHugh, *The Second Gilded Age: The Great Reaction in the United States, 1973–2001* (Lanham, MD: University Press of America, 2006). See also the classic Matthew Josephson, *The Robber Barons: The Great American Capitalists 1861–1901* (reprint) (New York: Harcourt, 2001).
25. Thomas Lemke, "Foucault, Governmentality, and Critique," *Rethinking Marxism* 14:3 (Fall 2002), pp. 49–64.
26. Nick Couldry, "Reality TV, or the Secret Theater of Neoliberalism," *Review of Education, Pedagogy, and Cultural Studies* 30:1 (January–March 2008), p. 1.
27. Brown, *Edgework*, p. 40.
28. Achille Mbembe, "Necropolitics," trans. Libby Meintjes, *Public Culture* 15:1 (2003), p. 40.
29. Kenneth Saltman and David Gabard, eds. *Education as Enforcement: The Militarization and Corporatization of Schools* (New York: Routledge, 2003).
30. Orlando Patterson, *Slavery and Social Death: A Comparative Study* (Cambridge, MA: Harvard University Press, 1982).
31. Chip Ward, "America Gone Wrong: A Slashed Safety Net Turns Libraries into Homeless Shelters," *TomDispatch.com* (April 2, 2007). Online: http://www.alternet.org/story/50023.
32. Ibid.

33. Faiz Shakir, Nico Pitney, Amanda Terkel, Satyam Khanna, and Matt Corley, "Katrina–FEMA Knowingly Allowed Katrina Refugees to Suffer from 'Toxic Trailers,'" *Progress Report* (July 20, 2007). Online: http://www.americanprogressaction.org/progressreport/2007/07bush_vetoes_kids.html. See also Amanda Spake, "Dying for a Home: Toxic Trailers Are Making Katrina Refugees Ill," *The Nation* (February 15, 2007). Online: http://www.alternet.org/module/printversion/48004.

34. Jean Comaroff, "Beyond Bare Life: AIDS, (Bio)Politics, and the Neoliberal Order," *Public Culture* 19:1 (Winter 2007), p. 213.

35. On the importance of the social state, see Zygmunt Bauman, *Liquid Fear* (London: Polity Press, 2006).

36. Paul Krugman, "Bush's Class-War Budget," *New York Times* (February 11, 2005). Online: http://www.nytimes.com/2005/02/11/opinion/11krugman.html?ex=1265864400&en=c5baff37424e2a5d&ei=5088&.

37. Faiz Shakir, Nico Pitney, Amanda Terkel, Satyam Khanna, and Matt Corley, "Bush Vetoes Kids," *The Progress Report* (July 20, 2007). Online: http://www.americanprogressaction.org/progressreport/2007/07/bush_vetoes_kids.html.

38. Ibid.

39. Peter Dreier's "Bush's Class Warfare" provides a detailed list of additional consequences arising from the class warfare waged by the former Bush administration:

> For example, Bush has handed the pharmaceutical industry windfall profits by restricting Medicare's ability to negotiate for lower prices for medicine. He targeted huge no-bid federal contracts to crony companies like Halliburton to supply emergency relief, reconstruction services and materials to rebuild Katrina while attempting to slash federal wage laws for reconstruction workers. He repealed Clinton-era "ergonomics" standards, affecting more than 100 million workers, that would have forced companies to alter their work stations, redesign their facilities or change their tools and equipment if employees suffered serious work-related injuries from repetitive motions. He opposed stiffer health and safety regulations to protect mine workers and cut the budget for federal agencies that enforce mine safety laws. Not surprisingly, under Bush, we've seen the largest number of mine accidents and deaths in years. Bush's Food and Drug Administration lowered product-labeling standards, allowing food makers to list health claims on labels before they have been scientifically proven. His FDA chief announced that the agency would no longer require claims to be based on "significant scientific agreement," a change that the National Food Processors Association, the trade association of the $500 billion food processing industry, had lobbied for. Bush resisted efforts to raise the minimum wage (which

had been stuck at $5.15 an hour for nine years) until the Democrats took back the Congress in 2006.

40. Marian Wright Edelman, "Where Are the Children in President Bush's Budget?" *Huffington Post* (April 8, 2008). Online: http://www.commondreams.org/archive/2008/03/10/7594/.

41. Ibid.

42. I take these issues up in detail in Henry A. Giroux, *The Abandoned Generation* (New York: Palgrave, 2004); and Henry A. Giroux and Susan Searls Giroux, *Take Back Higher Education* (New York: Palgrave, 2004).

43. John Dilulio, "The Coming of the Super-Predators," *Weekly Standard* 1:11 (November 27, 1995), p. 23. For an analysis of the drop in youth crime in the 1990s, see S. D. Levitt, "Understanding Why Crime Fell in the 1990s: Four Factors that Explain the Decline and Six that Do Not," *Journal of Economic Perspectives* 18:1 (Winter 2004), pp. 163–190.

44. Emily Gould, "Exposed," *New York Times* (May 28, 2008), pp. 32, 34–39, 52, 56.

45. I have taken this phrase from Garrett Keizer, "Requiem for the Private Word," *Harper's Magazine* (August 2008), p. 11.

46. Gould, "Exposed," p. 52.

47. Joel Walkowski, "Let's Not Get to Know Each Other Better," *New York Times* (June 8, 2008), p. ST6.

48. Ibid.

49. Madeleine Bunting, "From Buses to Blogs, a Pathological Individualism Is Poisoning Public Life," *The Guardian/UK* (January 28, 2008). Online: http://www.commondreams.org/archive/2008/01/28/6672/. For a critical examination of these issues, see two classic sources: Richard Sennett, *The Fall of Public Man* (New York: W.W. Norton, 1992); and Zygmunt Bauman, *Liquid Love: On the Frailty of Human Bonds* (London: Polity Press, 2003). See also Henry A. Giroux, *Public Spaces, Private Lives: Democracy Beyond 9/11* (Boulder: Rowman and Littlefield, 2003).

50. Judith Warner, "Kids Gone Wild," *New York Times* (November 27, 2005), p. A27.

51. Andy Newman, "In Trial, Defense Casts Slain Girl, 7, as a Terror," *New York Times* (January 17, 2008). Online: http://www.nytimes.com/2008/01/17/nyregion/17nixzmary.html.

52. Ibid.

53. Susan Searls Giroux, "Generation Kill: Nietzschean Meditations on the University, War, Youth, and Guns," *Works and Days* 51/52:1 and 2 (2008), p. 24.

54. Anya Kamentz, *Generation Debt: Why Now Is a Terrible Time to Be Young* (New York: Riverside, 2006), p. xiii.

55. Bob Herbert, "The Danger Zone," *New York Times* (March 15, 2007), p. A25.
56. Lawrence Grossberg, *Caught in the Crossfire* (Boulder: Paradigm Publishers, 2005), p. 16.
57. Bob Herbert, "Arrested While Grieving," *New York Times* (May 26, 2007), p. A25.
58. Lawrence Grossberg, "Why Does Neo-Liberalism Hate Kids? The War on Youth and the Culture of Politics," *Review of Education, Pedagogy, and Cultural Studies* 23:2 (2001), p. 133.
59. Howard Witt, "School Discipline Tougher on African Americans," *Chicago Tribune* (September 25, 2007). Online: http://www.chi cagotribune.com/news/nationworld/chi-070924discipline,0,22104. story?coll=chi_tab01_layout.
60. I have taken up a detailed critique of No Child Left Behind in Henry A. Giroux, *America on the Edge* (New York: Palgrave, 2006).
61. Steven Best and Douglas Kellner, "Contemporary Youth and the Postmodern Adventure," *Review of Education, Pedagogy, and Cultural Studies* 25:2 (April–June 2003), p. 78.
62. Lee Edelman has argued that claiming the child as a symbol of the future has largely shaped and been used by a politics that places enormous limits on the political field, reproducing what he calls "the privilege of heteronormativity." While this is no doubt partly true, it is not the only story about the use of the child as a figure either for referencing the political or for employing a discourse about the future that opens up rather than closes down democratic identities, structures, and possibilities. For instance, what are we to make of a right-wing politics that views the child as abject, other, and pathological precisely because children symbolize an element of social responsibility and a notion of the future at odds with the current massive attack on the social state, social investments, and any viable notion of the responsibility to future generations, if not the future itself? Moreover, suggesting that the figure of the child is crucial to understanding the increasing criminalization of young people and the ruthless workings of market fundamentalism in the United States does not assume that the child is the only organizing principle structuring a radical democratic politics. Surely, the discursive deployment of the child should not be rejected as merely a politics "based on a ponzi scheme of reproductive futurism." For two contrasting positions on this issue, see Lee Edelman, *No Future* (Durham: Duke University Press, 2004), p. 4; and Giroux, *The Abandoned Generation*.
63. American Academy of Pediatrics, "Prevention of Pediatric Overweight and Obesity," *Pediatrics* 112 (August 2003), pp. 424–430.
64. Naomi Klein, *No Logo* (New York: Picador, 1999), p. 177.

65. Bill Moyers, "A Time for Anger, A Call to Action," *Common Dreams* (February 7, 2007). Online: http://www.commondreams.org/views07/0322-24.htm.

66. See, especially, the invaluable political and educational work of Adam Fletcher and the vast array of youth groups he has brought together to address these issues. Adam Fletcher, *Youth Voice Handbook: The What, Who, Why, Where, When, and How Youth Voice Happens* (Olympia, WA: CommonAction, 2007). See also his work with The Free Child Project. Online: http://www.freechild.org/index.htm.

67. Jacques Derrida, "The Future of the Profession or the Unconditional University," in *Derrida Down Under*, ed. Laurence Simmons and Heather Worth (Auckland, New Zealand: Dunmarra Press, 2001), p. 253.

68. Zygmunt Bauman, *In Search of Politics* (Stanford: Stanford University Press, 1999), p. 2.

## CHAPTER 1

1. Jeremy Seabrook, *Consuming Cultures: Globalization and Local Lives* (Oxford: New Internationalist Publications, 2004), p. 270.

2. Stephen Kline, *Out of the Garden: Toys and Children's Culture in the Age of TV Marketing* (London: Verso, 1993), p. vii.

3. Lawrence Grossberg, *Caught in the Crossfire: Kids, Politics, and America's Future* (Boulder: Paradigm Publishers, 2005), p. 188.

4. The metaphor of the waste-disposal industry comes from the work of Zygmunt Bauman. The most recent work organized around this concept includes: Zygmunt Bauman, *Consuming Life* (London: Polity Press, 2007) and *Does Ethics Have a Chance in a World of Consumers* (Cambridge: Harvard University Press, 2008).

5. I have taken up this theme in a number of books. See Henry A. Giroux, *Disturbing Pleasures: Learning Popular Culture* (New York: Routledge, 1994); *Fugitive Cultures: Race, Violence, & Youth* (New York: Routledge, 1996); *Channel Surfing: Race Talk and the Destruction of Today's Youth* (New York: St. Martin's Press, 1997); *Stealing Innocence* (New York: Palgrave, 2000); *The Abandoned Generation: Democracy and the Culture of Fear* (New York: Palgrave Macmillan, 2003); and *America on the Edge* (New York: Palgrave Macmillan, 2006). See also Grossberg, *Caught in the Crossfire*.

6. Bauman, *Consuming Life*, p. 66.

7. Grossberg, *Caught in the Crossfire*.

8. See Michel Foucault, *The History of Sexuality: An Introduction* (New York: Vintage Books, 1990); *Society Must Be Defended: Lectures at the College De France 1975–1976* (New York: Picador, 1997); Michael

Hardt and Antonio Negri, *Empire* (Cambridge: Harvard University Press, 2000); *Multitude: War and Democracy in the Age of Empire* (New York: Penguin, 2004); and Giorgio Agamben, *Homo Sacer: Sovereign Power and Bare Life*, trans. Daniel Heller-Roazen (Stanford: Stanford University Press, 1998).

9. Mitchell Dean, "Four Theses on the Powers of Life and Death," *Contretemps* 5 (December 2004), p. 17.

10. Foucault, *Society Must Be Defended*, p. 249.

11. My discussion of waste and disposability in this chapter draws mainly on the work of Zygmunt Bauman, Mike Davis, Achille Mbembe, and Jeremy Seabrook. See Zygmunt Bauman, *Work, Consumerism and the New Poor* (Philadelphia: Open University Press, 1998); *The Individualized Society* (London: Polity Press, 2001); *Wasted Lives* (London: Polity Press, 2004); *Liquid Life* (London: Polity Press, 2005); *Liquid Fear* (London: Polity Press, 2006); *Liquid Times: Living in an Age of Uncertainty* (London: Polity Press, 2007); *Consuming Life*; and *Does Ethics Have a Chance*; Jeremy Seabrook, *Children of Other Worlds: Exploitation in the Global Market* (London: Pluto Press, 2001); Achille Mbembe, "Necropolitics," trans. Libby Meintjes, *Public Culture* 15:1 (2003), pp. 11–40; Seabrook, *Consuming Cultures*; Mike Davis, *Planet of Slums* (London: Verso, 2007); and Mike Davis and Daniel Bertrand Monk, *Evil Paradises: Dreamworlds of Neoliberalism* (New York: New Press, 2008). See also the excellent collection of articles in a special issue of *Cultural Studies* on anticonsumerism: Sam Binkley and Jo Littler, "Cultural Studies and Anti-Consumerism: A Critical Encounter," *Cultural Studies* 22:5 (September 2008).

12. As Lizabeth Cohen has brilliantly argued, the relationship between citizen and consumer over the course of the twentieth century defies any simple distinction and has been subject to ever-shifting categories. See Lizabeth Cohen, *A Consumer's Republic: The Politics of Mass Consumption in Postwar America* (New York: Vintage, 2003). This history seems undeniable to me. What I am suggesting is that under the biopolitics of neoliberalism, the complexity of this relationship has drastically narrowed, and any substantive notion of citizenship has become antithetical to dominant notions of consumerism.

13. Bauman, *Wasted Lives*, p. 6.

14. Ibid.

15. Jacques Derrida, "Autoimmunity: Real and Symbolic Suicides—A Dialogue with Jacques Derrida," in *Philosophy in a Time of Terror: Dialogues with Jurgen Habermas and Jacques Derrida*, ed. Giovanna Borradori (Chicago: University of Chicago Press, 2004), p. 94. For an extensive analysis and more favorable notion of immunity, see Roberto Esposito, *Bios: Biopolitics and Philosophy* (Minneapolis: University of Minnesota Press, 2008).

16. Susan Searls Giroux, "Generation Kill: Nietzschean Meditations on the University, War, Youth, and Guns," *Works and Days* 51/52:1 and 2 (2008), p. 1.

17. Jeremy Gilbert, "Against the Commodification of Everything: Anti-Consumerist Cultural Studies in the Age of Ecological Crisis," *Cultural Studies* 22:5 (September 2008), p. 53.

18. Bauman, *Consuming Life*, p. 12.

19. Grossberg, *Caught in the Crossfire*, p. 111.

20. Ibid., p. 117.

21. Bauman, *Liquid Life*, p. 62.

22. Wendy Brown, *Edgework: Critical Essays on Knowledge and Politics* (Princeton: Princeton University Press, 2005), p. 41.

23. For an excellent analysis of the control of corporate power on the media, see Robert W. McChesney, *The Political Economy of the Media* (New York: Monthly Review Press, 2008).

24. Bauman, *Liquid Life*, p. 114.

25. Colin Campbell, "I Shop Therefore I Know That I Am: The Metaphysical Basis of Modern Consumerism," in *Elusive Consumption*, ed. Karin M. Ekstrom and Henene Brembeck (New York: Berg, 2004), p. 42.

26. Patricia Cohen, "Energetic Rabbits, Melt-Proof Candies and Other Advertising Coups," *New York Times* (July 8, 2006), p. B6.

27. For an excellent history analyzing the politics of mass consumption in postwar America, see Lizabeth Cohen, *A Consumer's Republic*. For an equally impressive history of the commodification of childhood in the clothing industry, see Daniel Thomas Cook, *The Commodification of Childhood: The Children's Clothing Industry and the Rise of the Child Consumer* (Durham: Duke University Press, 2004). See also the now-classic work of Jackson Lears, *A Cultural History of Advertising in America* (New York: Basic Books, 1994).

28. Dan Cook, "Lunchbox Hegemony? Kids and the Marketplace, Then and Now," *LiP Magazine* (January 20, 2008). Online: http://www.lipmagazine.org/articles/featcook_124_p.html.

29. Brown, *Edgework*, p. 41.

30. Bauman, *Wasted Lives*, p. 70.

31. Chris Rojek, "The Consumerist Syndrome in Contemporary Society: An Interview with Zygmunt Bauman," *Journal of Consumer Culture* 4:3 (2004), p. 292.

32. See, for example, Bauman, *Liquid Times*; *Consuming Life*; and *Does Ethics Have a Chance*.

33. Bauman, *Liquid Life*, p. 68.

34. Jason Pine, "Economy of Speed," *Public Culture* 52 (Spring 2007), p. 358.

35. Rojek, "The Consumerist Syndrome," p. 301.

36. For two recent forays into this issue, see Russell Jacoby, *The Last Intellectuals* (New York: Basic Books, 2000) and Susan Jacoby, *The Age of*

*American Unreason* (New York: Pantheon, 2008). Of course, the classic work on anti-intellectualism is Richard Hofstadter, *Anti-Intellectualism in American Life* (New York: Vintage, 1966).

37. I have taken up this issue in great detail in Henry A. Giroux, *Public Spaces, Private Lives: Democracy Beyond 9/11* (Lanham, MD: Rowman and Littlefield, 2003).

38. Bauman cited in Rojek, "The Consumerist Syndrome," pp. 288–289.

39. Grossberg, *Caught in the Crossfire*, p. 117.

40. Bauman, *Wasted Lives*.

41. Bauman, *Consuming Life*, p. 21.

42. For a detailed treatment of liquid modernity, see Bauman, *Liquid Modernity* (London: Polity Press, 2000). See also Bauman, *Liquid Times*; and *Liquid Life*.

43. Both quotations are from Jonathan Rutherford, "Cultures of Capitalism: Seminar Paper for 23 November" *Soundings* 38 (Spring 2008). Online: http://www.lwbooks.co.uk/journals/soundings/cultures_capitalism/cultures_capitalism1.html

44. Bauman, *Liquid Life*, pp. 100–101.

45. See, for instance, the work of Angela Davis. In particular, see Angela Davis, "Recognizing Racism in the Era of Neoliberalism," Vice-Chancellor's Oration, delivered at Murdoch University (March 18, 2008). Online: www.abc.net.au/news/stories/2008/03/19/2193689.htm.

46. Bauman, *Wasted Lives*, p. 85.

47. Nancy Scheper-Hughes, "Bodies for Sale—Whole or In Parts," *Body & Society* 7:2–3 (2001), p. 2.

48. Herbert Marcuse, *One Dimensional Man* (Boston: Beacon Press, 1964); George Lukacs, *History and Class Consciousness* (Cambridge: MIT Press, 1972); Frederic Jameson, *Postmodernism, or the Cultural Logic of Late Capitalism* (London: Verso, 1991); Guy Debord, *The Society of the Spectacle*, trans. Donald Nicholson-Smith (New York: Zone Books, 1994); Jean Baudrillard, *Simulacra and Simulation* (Ann Arbor: University of Michigan Press, 1995); and Naomi Klein, *No Logo* (New York: Picador, 2000).

49. Some of the books that address the commercialization of young people while drawing on the theoretical legacy of now-classic theories of reification, consumption, and simulacra include: Jane Kenway and Elizabeth Bullen, *Consuming Children: Education–Entertainment–Advertising* (Philadelphia: Open University Press, 2001); Shirley Steinberg and Joe Kincheloe, *Kinderculture: The Corporate Construction of Childhood* (Boulder: Westview, 1997); Giroux, *The Abandoned Generation*; Grossberg, *Caught in the Crossfire*; Benjamin R. Barber, *Consumed: How Markets Corrupt Children, Infantilize Adults, and Swallow Citizens Whole* (New York: W. W. Norton, 2007); and Deron Boyles, *The Corporate Assault on Youth* (New York: Peter

Lang, 2008). Naomi Klein also focuses on young people who are exploited under horrendous labor conditions in order to produce the products that fill the demands of global neoliberalism as well as those who are the objects of ruthless marketing strategies. See Klein, *No Logo*.

50. Angela Y. Davis, *Abolition Democracy: Beyond Empire, Prisons, and Torture* (New York: Seven Stories Press, 2005), pp. 25–26.

51. Grossberg, *Caught in the Crossfire*, p. 264.

52. Barber, *Consumed*, p. 35.

53. See Josh Golin, "Nation's Strongest School Commercialism Bill Advances Out of Committee," *Common Dreams News Center* (August 1, 2007). Online: http://www.commondreams.org/cgi-bin/newsprint.cgi?file=/news2007/0801-06.htm. Juliet Schor argues that total advertising and marketing expenditures directed at children in 2004 reached $15 billion. See Juliet B. Schor, *Born to Buy* (New York: Scribner, 2005), p. 21.

54. Juliet Schor, "When Childhood Gets Commercialized Can Childhood Be Protected," in *Regulation, Awareness, Empowerment: Young People and Harmful Media Content in the Digital Age*, ed. Ulla Carlsson (Sweden: Nordicom, 2006), pp. 114–115.

55. Kiku Adatto, "Selling Out Childhood," *Hedgehog Review* 5:2 (Summer 2003), p. 40.

56. While the commodity form pervaded the social as early as the beginning of the nineteenth century, it not only has been normalized and extended into all aspects of social life in the latter half of the twentieth and beginning of the twenty-first century but also has legitimated a neoliberal politics of financialization that is far more ruthless in its destruction of the social and democracy than at any other time in the American past. On the history of consumption and commodification, see Lizabeth Cohen, *A Consumer's Republic*; and Lears, *A Cultural History of Advertising in America*. On neoliberalism and the politics of financialization, see David Harvey, *A Brief History of Neoliberalism* (New York: Oxford University Press, 2005); Wendy Brown, *Edgework*; Alfredo Saad-Filho and Deborah Johnston, eds., *Neoliberalism: A Critical Reader* (London: Pluto Press, 2005); Neil Smith, *The Endgame of Globalization* (New York: Routledge, 2005); Aihwa Ong, *Neoliberalism as Exception: Mutations in Citizenship and Sovereignty* (Durham: Duke University Press, 2006); Randy Martin, *An Empire of Indifference: American War and the Financial Logic of Risk Management* (Durham: Duke University Press, 2007); and Henry A. Giroux, *Against the Terror of Neoliberalism* (Boulder: Paradigm Publishers, 2008).

57. Schor, *Born to Buy*, p. 20.

58. Susan Linn, *Consuming Kids* (New York: Anchor Books, 2004), p. 8.

59. Barber, *Consumed*, pp. 7–8.

60. Schor, *Born to Buy*, p. 23.

61. Alex Molnar and Faith Boninger, "Adrift: Schools in a Total Marketing Environment," *Tenth Annual Report on Schoolhouse Commercialism Trends: 2006–2007* (Tempe: Arizona State University, 2007), pp. 6–7.

62. Anup Shah, "Children as Consumers," *Global Issues* (January 8, 2008). Online: http://www.globalissues.org/TradeRelated/Consumption/Children.asp.

63. Grossberg, *Caught in the Crossfire*, p. 88.

64. Linn, *Consuming Kids*, p. 54.

65. Molnar and Boninger, "Adrift," p. 9.

66. Schor, *Born to Buy*, pp. 19–20.

67. Cited in Brooks Barnes, "Web Playgrounds of the Very Young," *New York Times* (December 31, 2007). Online: http://www.nytimes.com/2007/12/31/business/31virtual.html?_r=1&oref=slogin.

68. Ibid.

69. Editorial, "Clothier Pushes Porn, Group Sex to Youths," *WorldNetDaily.com* (November 15, 2003). Online: http://www.wnd.com/news/article.asp?ARTICLE_ID=35604. See also Editorial, "Tell Nationwide Children's Hospital: No Naming Rights for Abercrombie & Fitch," *Campaign for a Commercial-Free Childhood* (June 2006). Online: http://salsa.democracyinaction.org/o/621/t/5401/campaign.jsp?campaign_KEY=23662.

70. Tana Ganeva, "Sexpot Virgins: The Media's Sexualization of Young Girls," *AlterNet* (May 24, 2008). Online: http://www.alternet.org/story/85977/.

71. Nick Turse, *How the Military Invades Our Everyday Lives* (New York: Metropolitan Books, 2008), p. 100.

72. Richard J. Bernstein, *The Abuse of Evil* (London: Polity Press, 2005), p. 49.

73. Douglas Kellner, *Guys and Guns Amok: Domestic Terrorism and the School Shootings from the Oklahoma City Bombing to the Virginia Tech Massacre* (Boulder: Paradigm, 2008), p. 157.

74. Victoria Rideout, Donald F. Roberts, and Ulla G. Foehr, *Generation M: Media in the Lives of 8–18 Year-Olds* (Washington, DC: Kaiser Family Foundation, March 2005), p. 4.

75. Ibid.

76. Jeff Chester and Kathryn Montgomery, *Interactive Food and Beverage Marketing: Targeting Children in the Digital Age* (Berkeley: Media Studies Group; Washington, DC: Center for Digital Democracy, 2007), p. 13. Online: http://digitalads.org/documents/digiMarketingFull.pdf.

77. Public time similarly differs from the fragmented model of time implied by news media, in which catastrophes exist in the time pocket of the news clip; each day, new catastrophes appear, only to be forgotten and replaced the next day with a different set of catastrophes. News stories are seldom followed long enough to suggest that events spring from complex causes and have equally complex repercussions, and the very

fact that stories are so quickly dropped implies that, no matter how catastrophic an event, its social importance is minimal.

78. Barber, *Consumed*, p. 231.

79. The Henry J. Kaiser Family Foundation, "The Role of Media in Childhood Obesity," (Menlo Park, CA: The Henry J. Kaiser Family Foundation, February 2004), p. 1. See also Zoe Williams, "Commercialization of Childhood," *Compass: Direction for the Democratic Left* (December 1, 2006). Online: http://www.criancaeconsumo.org. br/downloads/commercialization%20of%20childhood%20from %20britain.pdf. Williams estimates that children in both the United States and the United Kingdom are "exposed to between 20,000 and 40,000 ads a year."

80. Schor, *Born to Buy*, p. 25.

81. Rideout, Roberts, and Foehr, *Generation M*, pp. 6, 9.

82. Roger I. Simon, "On Public Time," Ontario Institute for Studies in Education. Unpublished paper, April 1, 2002, p. 4.

83. Barber, *Consumed*, p. 231.

84. For an excellent resource on youth activism, see Adam Fletcher, *Washington Youth Voice Handbook: The What, Who, Why, Where, When and How Youth Voice Happens* (Olympia, WA: CommonAction, 2007). Fletcher is one of the most important advocates for youth in the United States. See his Freechild Project Web site: http://www.freechild.org/index.htm. In Canada, one of the most important sources of criticism against neoliberalism and the commercialization of schools and the wider society can be found in the excellent work of Erika Shaker and the journal *Our Schools/Our Selves*. Sut Jhally in the United States through the Media Education Foundation has also waged an important pedagogical struggle through the popular media against neoliberalism and advertising. See Sut Jhally's excellent book *The Spectacle of Accumulation: Essays in Culture, Media, and Politics* (New York: Peter Lang, 2006).

85. Rutherford, "Cultures of Capitalism."

86. Schor, *Born to Buy*, pp. 16, 20.

87. Rutherford, "Cultures of Capitalism."

88. Juliet B. Schor, "The Commodification of Childhood: Tales from the Advertising Front Lines," *Hedgehog Review* 5:2 (Summer 2003), pp. 9–10.

89. A number of psychologists, especially Allen D. Kanner, have publicly criticized this practice by child psychologists. In fact, Kanner and some of his colleagues raised the issue in a letter to the American Psychological Association. See Miriam H. Zoll, "Psychologists Challenge Ethics of Marketing to Children," *American News Service* (April 5, 2000). Online: http://www.mediachannel. org/originals/kidsell.shtml. See also Allen D. Kanner, "The Corporatized Child," *California Psychologist* 39:1 (January/February 2006), pp. 1–2; and Allen D. Kanner,

"Globalization and the Commercialization of Childhood," *Tikkun* 20:5 (September/October, 2005), pp. 49–51. Kanner's articles are online: http://www.commercialfreechildhood.org/articles/.

90. Williams, "Commercialization of Childhood," p. 32.

91. Ibid., p. 34.

92. Madeleine Bunting, "In Our Angst over Children We're Ignoring the Perils of Adulthood," *The Guardian/UK* (November 13, 2006). Online: http://www.guardian.co.uk/commentisfree/2006/nov/13/comment.madeleinebunting.

93. I am borrowing here from Michel Foucault's notion of pastoral power. See Michel Foucault, *Security, Territory, Population: Lectures at the College de France, 1977–1978*, trans. Graham Burchell (New York: Palgrave Macmillan, 2007).

94. Michel Foucault, "Technologies of the Self," in *Essential Works of Foucault 1954–1984, Vol. 1: Ethics, Subjectivity and Truth*, ed. Paul Rabinow (Harmondsworth: Allen Lane/Penguin, 1997), p. 225.

95. Arjun Appadurai, *Fear of Small Numbers: An Essay on the Geography of Anger* (Durham: Duke University Press, 2006), p. 87.

96. I want to thank my dear colleague, David Clark, for this suggestion.

97. Molnar and Boninger, "Adrift," p. 1.

98. Carly Stasko and Trevor Norris, "Packaging Youth and Selling Tomorrow," in *The Corporate Assault on Youth*, ed. Deron Boyles (New York: Peter Lang, 2008), p. 130.

99. Molnar and Boninger provide a list of such companies along with their Web sites. See Molnar and Boninger, "Adrift," pp. 7–8.

100. See "About Us: The YMS Mission," (undated). Corporate Web site for YMS Consulting. Online: http://www.ymsconsulting.com/about_us.htm.

101. GIA Headquarters, "GIA Slumber Party in a Box." Online: http://www.giaheadquarters.com/sbox/faq.asp.

102. Bauman, *Wasted Lives*, p. 131.

103. Cited in Schor, *Born to Buy*, p. 65.

104. Bob Herbert, "Children in Peril," *New York Times* (April 21, 2009), p. A25.

105. Bauman, *Liquid Life*, p. 113.

106. The most notable example of this position is Neil Postman, *The Disappearance of Childhood* (New York: Vintage, 1994). I have taken up this issue in Henry A. Giroux, *The Mouse that Roared: Disney and the End of Innocence* (Boulder: Rowman and Littlefield, 1999); and *Stealing Innocence*. For a number of commentaries on the concept of childhood innocence, see Henry Jenkins, ed., *The Children's Culture Reader* (New York: New York University Press, 1998).

107. Tom Peters, "The Brand Called You," *Fast Company* (August/September 1997), p. 84.

108. Ibid., p. 86.

109. Ibid., p. 94.
110. Joseph E. Davis, "The Commodification of Self," *Hedgehog Review* 5:2 (Summer 2003), p. 41.
111. Lynn Hirschberg, "Bankable: How Tyra Banks Turned Herself Fiercely Into a Brand," *New York Times Sunday Magazine* (June 1, 2008), pp. 38–45, 58, 62–63.
112. Carlo Rotella, "And Now, the Biggest Entertainer in Entertainment," *New York Times* (June 1, 2008), p. 58.
113. On the matter of schools being modeled after businesses, see Kenneth Saltman, *Collateral Damage: Corporatizing Public Schools—A Threat to Democracy* (Boulder: Rowman and Littlefield, 2000); *The Edison Schools: Corporate Schooling and the Attack on Public Education* (New York: Routledge, 2005); Robin Goodman and Kenneth Saltman, *Strange Love: Or How We Learn to Stop Worrying and Love the Market* (Boulder: Rowman and Littlefield, 2002); Alex Molnar, *School Commercialism: From Democratic Ideal to Market Commodity* (New York: Routledge, 2005); and Molnar and Boninger, "Adrift."
114. Bauman, *Consuming Life*, pp. 91–92.
115. Bauman, "Happiness in a Society of Individuals," *Soundings* (Winter 2008), p. 23.
116. This argument is developed in Trevor Hogan, "The Space of Poverty: Zygmunt Bauman 'After' Jeremy Seabrook," *Thesis Eleven* 70 (August 2002), pp. 72–87.
117. Susan Gal, "Language, Gender, and Power: An Anthropological Review," in *Gender Articulated: Language and the Socially Constructed Self*, ed. Kira Hall and Mary Bucholtz (New York: Routledge, 1995), p. 178.
118. Steven Heller, *Iron Fists: Branding the 20$^{th}$-Century Totalitarian State* (New York: Phaidon Press, 2008).
119. Sheldon Wolin, *Democracy, Inc.: Managed Democracy and the Specter of Inverted Totalitarianism* (Princeton: Princeton University Press, 2008), pp. 12–13.
120. Stasko and Norris, "Packaging Youth," p. 131.
121. Mbembe, "Necropolitics," p. 11.
122. Kaiser Family Foundation Report, *The Role of the Media in Childhood Obesity* (Washington, DC: Kaiser Family Foundation, February 2004). Online: http://www.kff.org/entmedia/upload/The-Role-Of-Media-in-Childhood-Obesity.pdf.
123. Judith Warner, "Camp Codependence," *New York Times* (July 31, 2008). Online: http://warner.blogs.nytimes.com/2008/07/31/camp-codependence.
124. Claire Atkinson, "Kicked Out of Class: Primedia Sheds In-School Net Channel One," *Adage.com* (April 23, 2007). Online: http://www.commercialfreechildhood.org/news/primediashedschannelone.htm. For an excellent critical analysis, see Saltman, *The Edison Schools*.

125. Jerry Mander, "The Homogenization of Global Consciousness," *Lapis Magazine* (2001). Online: http://www.lapismagazine.org/index.php?option=com_content&task=view&id=120&Itemid=2.

126. Schor, "The Commodification of Childhood," p. 13.

127. Jeremy Seabrook, "Consuming Passions: Everything That Can Be Done to Bring the Age of Heroic Consumption to Its Close Should Be Done," *The Guardian/UK* (June 10, 2008). Online: http://www.guardian.co.uk/commentisfree/2008/jun/10/consumeraffairs.consumerandethicalliving.

128. Juliet Schor, "Tackling Turbo Consumption: An Interview with Juliet Schor," *Soundings* 34 (November 2006), p. 51.

129. Fulvia Carnevale and John Kelsey, "Art of the Possible: An Interview with Jacques Rancière," *Artforum* (March 2007), p. 264.

130. Wolin, *Democracy, Inc.*, pp. 260–261.

131. I have taken up this issue in a number of works; see especially Henry A. Giroux, *Against the New Authoritarianism: Politics After Abu Ghraib* (Winnipeg: Arbeiter Ring Publishing, 2005); and *Against the Terror of Neoliberalism*. See also Sheldon Wolin, *Democracy, Inc.*

132. Wolin, *Democracy, Inc.*, p. 261.

133. Bauman, *Liquid Life*, p. 14.

134. Sally Kohn, "Real Change Happens Off-Line," *Christian Science Monitor* (June 30, 2008). Online: http://www.csmonitor.com/2008/0630/p09s01-coop.html.

135. Jacques Derrida, "Intellectual Courage: An Interview," trans. Peter Krapp, in *Culture Machine* 2 (2000), p. 9.

## CHAPTER 2

1. Victor M. Rios, "The Hypercriminalization of Black and Latino Male Youth in the Era of Mass Incarceration," in *Racializing Justice, Disenfranchising Lives: The Racism, Criminal Justice, and Law Reader*, ed. Manning Marable, Ian Steinberg, and Keesha Middlemass (New York: Palgrave Macmillan), p. 17.

2. Richard J. Bernstein, *The Abuse of Evil: The Corruption of Politics and Religion Since 9/11* (London: Polity Press, 2005).

3. See, for example, James K. Galbraith, *The Predator State* (New York: Free Press, 2008).

4. David Theo Goldberg, *The Threat of Race* (Malden, MA: Wiley-Blackwell, 2009), p. 331.

5. Ibid., p. 331.

6. Glenn Greenwald, *How Would a Patriot Act? Defending American Values from a President Run Amok* (New York: Working Assets Publishing, 2006).

7. See the stinging critique and condemnation of the use of torture sanctioned by politicians at the highest level of the Bush administration in Editorial, "The Torture Report," *New York Times* (December 18, 2008), p. A34; see also Jane Mayer, *The Dark Side: The Inside Story of How the War on Terror Turned into a War on American Ideals* (New York: Doubleday, 2008).

8. The rise of the warfare state has been well documented. See Andrew Bacevich, *The Limits of Power* (New York: Metropolitan, 2008); Norman Solomon, *Made Love, Got War: Close Encounters with America's Warfare State* (New York: Polipoint Press, 2007); and Robert Higgs, *Resurgence of the Warfare State: The Crisis Since 9/11* (Washington, DC: The Independent Institute, 2005). On the increasing militarization of American society, see Henry A. Giroux, *The University in Chains: Confronting the Military-Industrial-Academic Complex* (Boulder: Paradigm, 2007); Chalmers Johnson, *Nemesis: The Last Days of the American Republic* (New York: Metropolitan, 2006); and Andrew Bacevich, *The New American Militarism: How Americans Are Seduced by War* (New York: Oxford University Press, 2005).

9. For an insightful commentary on the Bush administration's war against constitutional rights, see Greenwald, *How Would a Patriot Act?*; Philippe Sands, *Torture Team: Deception, Cruelty and the Compromise of Law* (New York: Allan Lane, 2008); and Mayer, *The Dark Side*.

10. Of course, there are some notable exceptions. See Jerome Miller, *Search and Destroy: African-American Males in the Criminal Justice System* (Cambridge: Cambridge University Press, 1996); David Cole, *No Equal Justice: Race and Class in the American Criminal Justice System* (New York: The New Press, 1999); Michael Parenti, *Lockdown America: Police and Prisons in the Age of Crisis* (London: Verso, 1999); Marc Mauer, *Race to Incarcerate* (New York: The New Press, 1999); Marc Mauer and Meda Chesney-Lind, *Invisible Punishment: The Collateral Consequences of Mass Imprisonment* (New York: The New Press, 2002); David Garland, *The Culture of Control: Crime and Social Order in Contemporary Society* (Chicago: University of Chicago Press, 2002); Mary Pattilo, David Weiman, and Bruce Western, eds, *Imprisoning America* (New York: Russell Sage Foundation, 2004); Angela Y. Davis, *Abolition Democracy: Beyond Empire, Prisons, and Torture* (New York: Seven Stories Press, 2005); Ruth Wilson Gilmore, *Golden Gulag: Prisons, Surplus, Crisis, and Opposition in Globalizing California* (Berkeley: University of California Press, 2007); Manning Marable, Ian Steinberg, and Keesha Middlemass, eds, *Racializing Justice, Disenfranchising Lives* (New York: Palgrave, 2007); and Bruce Western, *Punishment and Inequality in America* (New York: Russell Sage Foundation, 2007).

11. I take this up in Henry A. Giroux, *Fugitive Cultures: Race, Violence, and Youth* (New York: Routledge, 1996) and in *The Abandoned Generation* (New York: Palgrave, 2004).

12. Cited in Lawrence Grossberg, *Caught in the Crossfire: Kids, Politics, and America's Future* (Boulder: Paradigm Publishers, 2005), p. 4. The demonizing of kids has a long history, but it becomes intensified with the attack on the welfare state starting in the 1970s, the racist backlash that gains force under the Reagan administration in the 1980s, and the increased turn to criminalizing social problems under the Clinton and Bush administrations. For instance, kids were labeled "super-predators" in the 1990s; *Time* magazine in 1996 referred to five- and six-year-olds as "teenage time bombs"; and a Northeastern University criminologist added to the fire, claiming that young kids are "temporary sociopaths—impulsive and immature." See, for example, Richard Zoglin, "Now for the Bad News: A Teenage Time Bomb," *Time* (January 5, 1996). Online: http://www.time.com/time/magazine/article/0,9171,983959,00.html.

13. Linda S. Beres and Thomas D. Griffith, "Demonizing Youth," *Loyola of Los Angeles Law Review* 34 (January 2001), p. 747.

14. Jean Comaroff and John Comaroff, "Reflections of Youth, from the Past to the Postcolony," in *Frontiers of Capital: Ethnographic Reflections on the New Economy*, ed. Melissa S. Fisher and Greg Downey (Durham, NC: Duke University Press, 2006), p. 267.

15. Ibid.

16. Garland, *The Culture of Control*; and Jonathan Simon, *Governing Through Crime: How the War on Crime Transformed American Democracy and Created a Culture of Fear* (New York: Oxford University Press, 2007). See also Phil Scranton, *Power, Conflict and Criminalisation* (New York: Routledge, 2007).

17. Rios, "The Hypercriminalization of Black and Latino Male Youth," p. 17.

18. Jesse Hagopian, "The Dog Eats Its Tail: Oversized Classes, Overpopulated Prisons," *CommonDreams.Org* (March 7, 2009). Online: http://www.commondreams.org/view/2009/03/07-2.

19. Angela Y. Davis, "Recognizing Racism in the Era of Neoliberalism," Vice Chancellor's Oration (March 2008). Online: http://www.abc.net.au/news/opinion/speeches/files/20080318_davis.pdf.

20. For a brilliant analysis of the racist state, see David Theo Goldberg, *The Racial State* (Malden, MA: Wiley-Blackwell, 2001).

21. Joe Klein, "Obama's Victory Ushers in a New America," *Time.com* (November 5, 2008). Online: http://www.time.com/time/politics/article/0,8599,1856649,00.html.

22. Paul Ortiz, "On the Shoulders of Giants: Senator Obama and the Future of American Politics," *Truthout.org* (November 25, 2008). Online: http://www.truthout.org/112508R?print.

23. Simon, *Governing Through Crime*, p. 59.

24. Davis, *Abolition Democracy*, p. 98.

25. From a transcript entitled "Barack Obama's Speech on Race," *New York Times* (March 18, 2008). Online: http://www.nytimes.com/

2008/03/18/us/politics/18text-obama.html?_r=1&scp=1&sq=%22Barack%20Obama's%20Speech%20on%20Race%22&st=cse.

26. Zygmunt Bauman, "Happiness in a Society of Individuals," *Soundings* (Winter 2008), pp. 22–23.

27. Zygmunt Bauman, *Consuming Life* (London: Polity Press, 2007), pp. 126, 128.

28. This issue is taken up in a number of important books. See Christopher Robbins, *Expelling Hope* (Albany: SUNY Press, 2008); Valerie Polakow, *Who Cares for Our Children* (New York: Teachers College Press, 2007); William Lyons and Julie Drew, *Punishing Schools: Fear and Citizenship in American Public Education* (Ann Arbor: University of Michigan Press, 2006); Giroux, *The Abandoned Generation*; Valerie Polakow, ed., *The Public Assault on America's Children: Poverty, Violence, and Juvenile Justice* (New York: Teachers College Press, 2000); and William Ayers, *A Kind and Just Parent* (Boston: Beacon Press, 1997).

29. I have argued that the prison-industrial complex has to be analyzed within the rise of a new mode of authoritarianism that emerged under the Bush-Cheney administration. See Henry A. Giroux, *Against the Terror of Neoliberalism* (Boulder: Paradigm Publishers, 2008). See also Sheldon S. Wolin, *Democracy Inc.: Managed Democracy and the Specter of Inverted Totalitarianism* (Princeton: Princeton University Press, 2008).

30. Davis, *Abolition Democracy*, pp. 117, 40–41.

31. On the issue of the racial biopolitics of neoliberalism, war, violence, authoritarianism, and everyday life, see Leerom Medovoi, "Global Society Must Be Defended: Biopolitics Without Boundaries," *Social Text* 25:2 (Summer 2007), pp. 53–79; Jonathan Michael Feldman, "From Warfare State to 'Shadow State,'" *Social Text* 25:2 (Summer 2007), pp. 143–168; James M. Cypher, "From Military Keynesianism to Global-Neoliberal Militarism," *Monthly Review* (June 2007), pp. 1–18; Nikhil Singh, "The Afterlife of Fascim," *South Atlantic Quarterly* 105:1 (Winter 2006), pp. 71–93; and Stephen John Hartnett and Laura Ann Stengrim, "War Rhetorics," *South Atlantic Quarterly* 105:1 (Winter 2006), pp. 175–205. For an interesting commentary on how the logic of war and its politics of disposability render human beings into machines, see the brilliant essay by David Clark, "Schelling's Wartime: Philosophy and Violence in the Age of Napoleon," *European Romantic Review* 19:2 (April 2008), pp. 139–148.

32. Michel Foucault, *The Birth of Biopolitics: Lectures at the College de France, 1978–1979* (New York: Palgrave, 2008), p. 323.

33. Zygmunt Bauman, *Wasted Lives* (London: Polity Press, 2004), p. 6.

34. Rios, "The Hypercriminalization of Black and Latino Male Youth," p. 21.

35. Ibid., pp. 17–18.

36. This argument is made brilliantly in Jonathan Simon, *Governing Through Crime*. I rely heavily on Simon's work in this section of the chapter, though I extend his model of crime to the politics of disposability. On the politics of disposability, I am heavily indebted to the wide-ranging work of Zygmunt Bauman, cited throughout this chapter and book.
37. Goldberg, *The Threat of Race*, p. 335.
38. Simon, *Governing Through Crime*, p. 4.
39. Ibid., pp. 15–16.
40. Ibid., p. 5.
41. Davis, *Abolition Democracy*, p. 41.
42. Simon, *Governing Through Crime*, p. 14.
43. Rios, "The Hypercriminalization of Black and Latino Male Youth," p. 21.
44. Gilmore, *Golden Gulag*, p. 242.
45. Bauman, *Consuming Life*, pp. 132–133.
46. Simon, *Governing Through Crime*, pp. 142–143.
47. Davis, *Abolition Democracy*, p. 113.
48. Gilmore, *Golden Gulag*, p. 5.
49. Simon, *Governing Through Crime*, p. 19.
50. See, for example, Michael Hardt and Antonio Negri, *Multitude: War and Democracy in the Age of Empire* (New York: Penguin, 2005). I have also taken up this issue in Giroux, *The University in Chains*; and Giroux, *Against the Terror of Neoliberalism*.
51. Jennifer Warren, *One in 100: Behind Bars in America 2008* (Washington, DC: The PEW Center on the States, 2007), pp. 3, 5.
52. Davis, "Recognizing Racism in the Era of Neoliberalism."
53. Warren, *One in 100: Behind Bars in America 2008*, pp. 5, 7.
54. Jason DeParle, "The American Prison Nightmare," *New York Review of Books* 54:6 (April 12, 2007), p. 33.
55. Paul Street, *Segregated Schools: Educational Apartheid in Post-Civil Rights America* (New York: Routledge, 2005), p. 82.
56. Loic Wacquant, "From Slavery to Mass Incarceration: Rethinking the 'Race Question' in the U.S," *New Left Review* (January–February 2002), pp. 56–57.
57. Erik Eckholm, "Reports Find Racial Gap in Drug Arrests," *New York Times* (May 5, 2008), p. A21.
58. Sam Dillon, "Hard Times Hitting Students and Schools," *New York Times* (September 1, 2008), pp. A1, A9.
59. Warren, *One in 100: Behind Bars in America 2008*, pp. 4, 11.
60. Pew Center on the States, *One in 31: The Long Reach of American Corrections* (Washington, DC: The Pew Charitable Trusts, March 2009). p. 11.
61. Adam Liptak, "1 in 100 U.S. Adults Behind Bars, New Study Says," *New York Times* (February 28, 2008). Online: http://www.nytimes.

com/2008/02/28/us/28cnd-prison.html?sq=Adam%20Liptak&st=cse&scp=34&pagewanted=print.

62. James Sterngold, "Prisons' Budget to Trump Colleges': No Other Big State Spends as Much to Incarcerate Compared with Higher Education Funding," *San Francisco Chronicle* (May 21, 2007). Online: http://www.sfgate.com/cgi-bin/article.cgi?f=/c/a/2007/05/21/MNG4KPUKV51.DTL.

63. Wacquant, "From Slavery to Mass Incarceration," p. 44.

64. DeParle, "The American Prison Nightmare," p. 33. For an extensive treatment of how mass incarceration plays a major role in producing inequality, see Western, *Punishment and Inequality in America*.

65. Brent Staples, "Growing Up in the Visiting Room," *New York Times Book Review* (March 21, 2004). Online: http://query.nytimes.com/gst/fullpage.html?res=9A07EFD6123EF932A15750C0A9629C8B63.

66. DeParle, "The American Prison Nightmare," pp. 35–36.

67. Ibid., pp. 33–34.

68. Arianna Huffington, "The Criminal Justice Reform Battle in California: Cynical Politicians and Powerful Interests Attacking the Public Good," *AlterNet* (November 2, 2008). Online: http://www.alternet.org/story/105741/.

69. Daniel Macallair and G. Thomas Gitchoff, "California's Big Chance to Stop Locking Up Harmless Drug Offenders," *AlterNet* (October 25, 2008). Online: http://www.alternet.org/story/104589/.

70. Adam Nossiter, "With Jobs to Do, Louisiana Parish Turns to Inmates," *New York Times* (July 5, 2006). Online: http://www.nytimes.com/2006/07/05/us/05prisoners.html.

71. Ibid.

72. Under such conditions, as Derrida points out, the Arab became the "ultimate figure of exclusion and dissidence in the post-9/11 era." See Giovanna Borradori, "Foreword," in Mustapha Cherif, *Islam and the West: A Conversation with Jacques Derrida* (Chicago: University of Chicago Press, 2008), p. x.

73. Bauman, *Consuming Life*, p. 128.

74. Grossberg, *Caught in the Crossfire*, p. 36.

75. "Generation Kill" is the name of a seven-part HBO television miniseries about what the *New York Times* calls "a group of shamelessly and engagingly profane, coarse and irreverent marines . . . that spearhead the invasion" in the second Iraq war. See Alessandra Stanley, "Comrades in Chaos, Invading Iraq," *New York Times* (July 11, 2008), p. B1. The term "Killer Children" appears as the title of a *New York Times* book review. See Kathryn Harrison, "Killer Children," *New York Times Book Review* (July 20, 2008), pp. 1, 8.

76. Douglas Kellner, *Guys and Guns Amok: Domestic Terrorism and School Shootings from the Oklahoma City Bombing to the Virginia Tech Massacre* (Boulder: Paradigm Publishers, 2008), p. 14.

77. Russ Bynum, "Gr. 3 Kids Plotted to Stab Teacher, Police Say," *Toronto Star* (April 2, 2008). Online: http://www.thestar.com/printArticle/ 408969.

78. On the dark side of the war on terror and the politics of torture, see Philippe Sands, *Torture Team: Deception, Cruelty and the Compromise of Law* (New York: Allan Lane, 2008); and Jane Mayer, *The Dark Side.*

79. Bob Herbert, "Out of Sight," *New York Times* (June 10, 2008), p. A23.

80. Bruce E. Levine, "How Teenage Rebellion Has Become a Mental Illness," *AlterNet* (January 28, 2008), online: http://www. alternet.org/story/75081; and Benedict Carey, "Use of Antipsychotics by the Young Rose Fivefold," *New York Times* (June 6, 2006), online: http://www.nytimes.com/2006/06/06/health/06psych. html?_r=1& scp=1&sq=%93Use%20of%20Antipsychotics%20by%20the%20Young %20Rose%20Fivefold,%94%20&st=cse&oref=slogin.

81. Lindsay Beyerstein Majikthise, "Congress Probes Child Abuse Under Guise of Treatment," *AlterNet* (May 7, 2008). Online: http://www.alternet.org/bloggers/majikthise.typead.com/84572/. For an in-depth analysis of the tough-love industry aimed at children, see Maia Szalavitz, *Help at Any Cost: How the Troubled-Teen Industry Cons Parents and Hurts Kids* (New York: Riverside, 2006).

82. Alex Koroknay-Palicz, "Scapegoating of Youth," *National Youth Rights Association* (December 2001). Online: http://www.youthrights.org/ scapegoat.php.

83. Children's Defense Fund, *2008 Annual Report* (Washington, DC: Children's Defense Fund, 2009). Online: http://www. childrensdefense. org/child-research-data-publications/data/state-of-americas-children-2008-report.pdf.

84. See Bob Herbert, "Head for the High Road," *New York Times* (September 2, 2008), p. A25; Sam Dillon, "Hard Times Hitting Students and Schools," *New York Times* (September 1, 2008), pp. A1, A9; and Erik Eckholm, "Working Poor and Young Hit Hard in Downturn," *New York Times* (November 9, 2008), p. A23.

85. For an excellent analysis of this issue, see Robbins, *Expelling Hope.* See also Lyons and Drew, *Punishing Schools*; and Giroux, *The Abandoned Generation.*

86. Simon, *Governing Through Crime,* p. 209.

87. Ian Urbina and Sean D. Hamill, "Judges Plead Guilty in Scheme to Jail Youths for Profit," *New York Times* (February 13, 2009), p. A1.

88. Ibid.

89. Ibid.

90. Bob Herbert, "School to Prison Pipeline," *New York Times* (June 9, 2007), p. A29.

91. Ibid.

92. Randall R. Beger, "Expansion of Police Power in Public Schools and the Vanishing Rights of Students," *Social Justice* 29:1 (2002), p. 120.

93. This term comes from Garland, *The Culture of Control*.
94. The best book analyzing all aspects of zero tolerance policies is Robbins, *Expelling Hope*. See also William Ayers, Bernardine Dohrn, and Rick Ayers, eds., *Zero Tolerance* (New York: The New Press, 2001). I have also taken up this issue in Giroux, *The Abandoned Generation*.
95. Yolanne Almanzar, "First Grader in $1 Robbery May Face Expulsion," *New York Times* (December 4, 2008), p. A26.
96. Advancement Project in partnership with Padres and Jovenes Unidos, Southwest Youth Collaborative, *Education on Lockdown: The Schoolhouse to Jailhouse Track* (Chicago: Children and Family Justice Center of Northwestern University School of Law, March 24, 2005), p. 11.
97. Ibid., p. 33.
98. Ibid., p. 7.
99. Bernardine Dohrn, " 'Look Out, Kid, It's Something You Did': The Criminalization of Children," in *The Public Assault on America's Children*, ed. Valerie Polakow, p. 158.
100. Advancement Project, *Education on Lockdown*, p. 18.
101. Ibid., pp. 17–18.
102. Ibid., p. 31.
103. Elora Mukherjee, *Criminalizing the Classroom: The Over-Policing of New York City Schools* (New York: American Civil Liberties Union and New York Civil Liberties, March 2008), p. 9.
104. Ibid., p. 6.
105. Ibid., p. 16.
106. Ibid., p. 16.
107. Beger, "Expansion of Police Power," p. 120.
108. Rios, "The Hypercriminalization of Black and Latino Male Youth," pp. 40–54.
109. For a superb analysis of urban marginality of youth in the United States and France, see Loic Wacquant, *Urban Outcasts* (London: Polity Press, 2008).
110. Children's Defense Fund, *America's Cradle to Prison Pipeline* (Washington, DC: Children's Defense Fund, 2008), p. 77. Online: http://www.childrensdefense.org/site/DocServer/CDF_annual_report_07.pdf?docID=8421.
111. Elizabeth White, "The Latest Back-to-School Fashion: Prison-style Jumpsuits," *The Globe and Mail* (August 4, 2008), p. L4.
112. Associated Press, "Rebels Turn Texas School Jumpsuit Plan Inside Out," *ABC News* (August 5, 2008). Online: http://abcnews.go.com/US/wireStory?id=5498827.
113. White, "The Latest Back-to-School Fashion," p. L4.
114. Dohrn, " 'Look Out, Kid, It's Something You Did'," p. 175.
115. For a moving and insightful book on incarcerating kids, see William Ayers, *A Kind and Just Parent* (Boston: Beacon Press, 1997). On the issue of juvenile justice reform, see the Annie E. Casey Foundation, "A

Road Map for Juvenile Justice Reform," in *2008 Kids Count Data Book* (Baltimore: Annie E. Casey Foundation, 2008). Online: http://www. aecf.org/~/media/PublicationFiles/AEC180essay_booklet_MECH. pdf. On the issue of juvenile justice and the cradle-to-prison pipeline, see Children's Defense Fund, *America's Cradle to Prison Pipeline*; Equal Justice Initiative, *Cruel and Unusual: Sentencing 13- and 14-Year-Old Children to Die in Prison*, online: http://www.eji.org/eji/files/ 20071017cruelandunusual.pdf; and A Campaign for Youth Justice Report, *Jailing Juveniles: The Dangers of Incarcerating Youth in Adult Jails in America* (Washington, DC: Campaign for Youth Justice, November 2007), online: http://www.campaignforyouthjustice.org/national_reports.html. See also David Tanenhaus, *Juvenile Justice in the Making* (New York: Oxford University Press, 2005).

116. Annie E. Casey Foundation, "A Road Map for Juvenile Justice Reform."

117. Equal Justice Initiative, *Cruel and Unusual*.

118. Ibid.

119. Ibid.

120. Marian Wright Edelman, "Juveniles Don't Belong in Adult Prisons," *Children's Defense Fund* (August 1, 2008). Online: http://www. huffingtonpost.com/marian-wright-edelman/juveniles-don't-belong-in_b_116747.html.

121. Ibid.

122. Equal Justice Initiative, *Cruel and Unusual*.

123. Ibid.

124. Ibid.

125. Ibid.

126. Ibid.

127. Grossberg, *Caught in the Crossfire*, p. 308.

128. We take up this issue in more detail in Henry A. Giroux and Kenneth Saltman, "Obama's Betrayal of Public Education? Arne Duncan and the Corporate Model of Schooling," *Truthout.org* (December 17, 2008). Online: http://www.truthout.org/121708R?print. I also want to thank a teacher, Paul A. Moore, for his personal correspondence on this issue, dated December 18, 2008.

## CHAPTER 3

1. Zygmunt Bauman, *The Individualized Society* (London: Polity Press, 2001), pp. 54–55.

2. Anya Kamenetz, *Generation Debt* (New York: Riverhead, 2006); and Barbara Ehrenreich, "College Students, Welcome to a Lifetime of Debt?" *AlterNet* (September 11, 2007), online: http://www.alternet. org/workplace/62125/.

3. Frank Donoghue, The Last Professor: The Corporate University and the Fate of the Humanities (New York: Fordham University Press, 2008).

4. Tamar Lewin, "College May Become Unaffordable for Most in U.S.," New York Times (December 3, 2008), p. A17.

5. David Moltz, "Tenure on the Chopping Block," Inside Higher Ed (December 3, 2008). Online: http://www.insidehighered.com/news/2008/12/03/kentucky. Of course, there is a long and venerable tradition that is critical of the university as an adjunct of business interests, power, and culture. One viable starting point would be Thorstein Veblen, The Higher Learning in America: A Memorandum on the Conduct of Universities by Businessmen (New York: Augustus M. Kelley, 1965).

6. A. Lee Fritschler, Bruce L. R. Smith, and Jeremy D. Mayer, Closed Minds? Politics and Ideology in American Universities (Washington, DC: Brookings Institution, 2008).

7. I have taken up this issue in detail in Henry A. Giroux and Susan Searls Giroux, Take Back Higher Education (New York: Palgrave, 2004).

8. Doug Henwood, After the New Economy (New York: The New Press, 2005).

9. I discuss this phenomenon in Henry A. Giroux, The University in Chains: Confronting the Military-Industrial-Academic Complex (Boulder: Paradigm, 2007).

10. Alan Finder, "Decline of the Tenure Track Raises Concerns," New York Times (November 20, 2007), p. A16.

11. I take these issues up in great detail in Giroux, The University in Chains.

12. Ian Angus, "Academic Freedom in the Corporate University," in Utopian Pedagogy: Radical Experiments Against Neoliberal Globalization, ed. Mark Coté, Richard J. F. Day, and Greig de Peuter (Toronto: University of Toronto Press, 2007), p. 69.

13. Philip Leopold, "The Professorial Entrepreneur," Chronicle of Higher Education (August 30, 2007). Online: http://chronicle.com/jobs/news/2007/08/2007083001c/careers.html.

14. Jeffrey Brainard, "U.S. Defense Secretary Asks Universities for New Cooperation," Chronicle of Higher Education (April 16, 2008). Online: http://chronicle.com/news/article/4316/ us-defense-secretary-asks-universities-for-new-cooperation.

15. I discuss scholarship programs linked to the CIA in detail in Giroux, The University in Chains, pp. 62–72.

16. Henry A. Giroux and Kenneth Saltman, "Obama's Betrayal of Public Education? Arne Duncan and the Corporate Model of Schooling, Truthout.org (December 17, 2008). Online: http://www.truthout.org/121708R.

17. Jennifer Washburn, University, Inc.: The Corporate Corruption of Higher Education (New York: Basic Books, 2006), p. 227.

18. I take this issue up in great detail in Giroux and Giroux, *Take Back Higher Education*; and Henry A. Giroux, *Against the Terror of Neoliberalism* (Boulder: Paradigm, 2008).

19. John Dewey cited in E. L. Hollander and John Saltmarsh, "The Engaged University," *Academe* (July/August, 2000). Online: http://www.aaup.org/AAUP/pabsres/academe/2000/JA/holl.htm.

20. Wendy Brown, *Regulating Aversion* (Princeton: Princeton University Press, 2006), p. 88.

21. John Dewey, *Individualism: Old and New* (New York: Minton, Balch, 1930), p. 41.

22. Richard J. Bernstein, *The Abuse of Evil: The Corruption of Politics and Religion Since 9/11* (London: Polity Press, 2005), p. 45.

23. James B. Conant, "Wanted: American Radicals," *The Atlantic* (May 1943). Online: http://www.theatlantic.com/issues/95sep/ets/radical.htm.

24. Chris Hedges, "America Is in Need of a Moral Bailout," *Truthdig.com* (March 23, 2009). Online: http://www.truthdig.com/report/item/20090323_america_is_in_need_of_a_moral_bailout/.

25. Doug Lederman, "Rethinking Student Aid, Radically," *Inside Higher Ed* (September 19, 2008). Online: http://www.insidehighered.com/news/2008/09/19/rethink.

26. For an excellent analysis of this attack, see Beshara Doumani, "Between Coercion and Privatization: Academic Freedom in the Twenty-First Century," in *Academic Freedom After September 11*, ed. Doumani (Cambridge, MA: Zone Books, 2006), pp. 11–57; and Evan Gerstmann and Matthew J. Streb, *Academic Freedom at the Dawn of a New Century: How Terrorism, Governments, and Culture Wars Impact Free Speech* (Stanford: Stanford University Press, 2006). A sustained and informative discussion of academic freedom after 9/11 can be found in Tom Abowd, Fida Adely, Lori Allen, Laura Bier, and Amahl Bishara et al., *Academic Freedom and Professional Responsibility After 9/11: A Handbook for Scholars and Teachers* (New York: Task Force on Middle East Anthropology, 2006), online: http://www.meanthro.org/Handbook-1.pdf. See also AAUP, "Academic Freedom and National Security in a Time of Crisis," *Academe* 89:6 (2003), online: http://www. aaup.org/AAUP/About/committees/committee+repts/cristime.htm; Jonathan R. Cole, "Academic Freedom Under Fire," *Daedalus* 134:2 (2005), pp. 1–23; American Federation of Teachers, *Academic Freedom in the 21st-Century College and University* (2007), online: http://www.aft.org/pubs-reports/higher-ed/AcademicFreedomStatement.pdf; and American Association of University Professors, "Freedom in the Classroom," *AAUP Report* (September-October, 2007), online: http://www.aaup.org/AAUP/comm/rep/A/class.htm.

27. Editorial, "Targeting the Academy," *Media Transparency* (March 2003). Online: http://www.mediatransparency.org/conservative philanthropy.php?conservativePhilanthropyPageID=11.
28. Ibid.
29. Max Blumenthal, "Princeton Tilts Right," *The Nation* (March 13, 2006), p. 14.
30. Kelly Field, "Recruiting for the Right," *Chronicle of Higher Education* (January 12, 2007), p. A35.
31. Patricia Cohen, "Conservatives Try New Tack on Campuses," *New York Times* (September 22, 2008), p. A22.
32. Lewis F. Powell, Jr., "The Powell Memo," *ReclaimDemocracy.org* (August 23, 1971). Online: http://reclaimdemocracy.org/corporate_accountability/powell_memo_lewis.html.
33. Ibid.
34. Lewis H. Lapham, "Tentacles of Rage—The Republican Propaganda Mill, a Brief History," *Harper's Magazine* (September 2004), p. 32.
35. Dave Johnson, "Who's Behind the Attack on Liberal Professors?" *History News Network* (February 10, 2005). Online: http://hnn.us/articles/printfriendly/1244.html.
36. Alan Jones, "Connecting the Dots," *Inside Higher Ed* (June 16, 2006). Online: http://insidehighered.com/views/2006/06/16/jones.
37. Ellen Schrecker, "Worse Than McCarthy," *Chronicle of Higher Education* 52:23 (February 10, 2006), p. B20.
38. Joel Beinin, "The New McCarthyism: Policing Thought about the Middle East," in *Academic Freedom after September 11*, ed. Beshara Doumani (New York: Zone Books, 2006), p. 242.
39. Jerry L. Martin and Anne D. Neal, *Defending Civilization: How Our Universities Are Failing America and What Can Be Done About It*, ACTA Report (November 2001). Online: http://www.la.utexas.edu/~chenry/2001LynnCheneyjsg01ax1.pdf. This statement was deleted from the revised February 2002 version of the report available on the ACTA Web site: http://www.goacta.org/publications/Reports/defciv.pdf.
40. I have taken this term, at least part of it, from a quotation by Sheila Slaughter. Cited in Richard Byrne, "Scholars See Need to Redefine and Protect Academic Freedom," *Chronicle of Higher Education* (April 7, 2008). Online: http://chronicle.com/daily/2008/04/2384n.htm.
41. See http://www.targetofopportunity.com/enemy_targets.htm.
42. Nicholas Turse, "The Military-Academic Complex," *TomDispatch.com* (April 29, 2004). Online: http://www.countercurrents.org/us-turse 290404.htm.
43. I take these cases up in great detail in Henry A. Giroux, "Academic Freedom Under Fire: The Case for Critical Pedagogy," *College Literature* 33:4 (Fall 2006), pp. 1–42.

44. Jonathan R. Cole, "The New McCarthyism," *Chronicle of Higher Education* 52:3 (September 9, 2005), p. B7.

45. American Council of Trustees and Alumni, *How Many Ward Churchills?: A Study by the American Council of Trustees and Alumni* (Washington, DC: American Council of Trustees and Alumni, May 2006), p. 22.

46. Ibid., p. 2.

47. Ellen Schrecker cited in Justin M. Park, "Under Attack: Free Speech on Campus," *Clamor* 34 (September/October, 2005). Online: http://clamormagazine.org/isues/34/culture.php.

48. ACTA, *How Many Ward Churchills?*, p. 12.

49. James Pierson, "The Left University," *Weekly Standard* 11:3 (October 3, 2005). Online: http://www.weeklystandard.com/Content/Public/Articles/000/000/006/120xbklj.asp.

50. Roger Kimball, "Rethinking the University: A Battle Plan," *The New Criterion* 23 (May 2005). Online: http://www.newcriterion.com/archive/23/may05/universe.htm.

51. CBN News, transcript of an interview with David Horowitz, "The 101 Most Dangerous Professors in America," *CBN News.com* (March 22, 2006). Online: http://cbn.com/cbnnews/commentary/060322a.aspx.

52. Cole, "Academic Freedom Under Fire."

53. Cited in Jennifer Jacobson, "What Makes David Run," *Chronicle of Higher Education* (May 6, 2005), p. A9.

54. The Academic Bill of Rights is available online: http://www.studentsforacademicfreedom.org/abor.html.

55. David Horowitz, "In Defense of Intellectual Diversity," *Chronicle of Higher Education* 50:23 (February 13, 2004), p. B12.

56. See, for instance, John K. Wilson, *Patriotic Correctness: Academic Freedom and Its Enemies* (Boulder: Paradigm, 2006); Russell Jacoby, "The New PC: Crybaby Conservatives," *The Nation* (April 4, 2006), pp. 11–15; Martin Plissner, "Flunking Statistics: The Right's Misinformation about Faculty Bias," *American Prospect* 13 (December 30, 2002), online: http://www.prospect.org/cs/articles?article=flunking_statistics; and Yoshie Furuhashi, "Conservatives: Underrepresented in Academia?" *Critical Montages* (April 2, 2005), online: http://montages.blogspot.com/2005/04/conservatives-underrepresented-in.html.

57. See Lionel Lewis's response to Anne D. Neal in "Political Bias on Campus," *Academe* (May 5, 2005). Online: http://www.aaup.org/publications/Academe/2005/05/mj/05mjlte.htm.

58. Gary Younge, "Silence in Class," *The Guardian* (April 3, 2006). Online: http://www.guardian.co.uk/usa/story/0,1746227,00. html.

59. Jennifer Jacobson, "Conservatives in a Liberal Landscape," *Chronicle of Higher Education* 51:5 (September 24, 2004), pp. A8–A11.

60. John F. Zipp and Ruddy Fenwick, "Is the Academy a Liberal Hegemony? The Political Orientation and Educational Values of Professors," *Public Opinion Quarterly* 70:3 (2006). Online: http://poq.oxfordjournals.org/cgi/content/full/70/3/304?ijkey=dVt13UcYfsj5AyF& keytype=ref#SEC5.

61. Stanley Fish, "On Balance," *Chronicle of Higher Education* (April 1, 2005). Online: http://chronicle.com/jobs/2005/04/ 2005040101c. htm.

62. Russell Jacoby, "The New PC," p. 13.

63. The Students for Academic Freedom Web site: http://www.studentsforacademicfreedom.org.

64. "SAF Complaint Center" Web site: http://www. studentsforacademicfreedom.org/comp/default.asp.

65. Robert Ivie, "Academic Freedom and Political Heresy," *IU Progressive Faculty Coalition Forum* (October 3, 2005). Online: http://www.indiana.edu/~ivieweb/academicfreedom.htm.

66. Stuart Hall, "Epilogue: Through the Prism of an Intellectual Life," in *Culture, Politics, Race and Diaspora: The Thought of Stuart Hall,* ed. Brian Meeks (London: Lawrence and Wishart, 2007), p. 270.

67. Fulvia Carnevale and John Kelsey, "Art of the Possible: An Interview with Jacques Rancière," *Artform* (March 2007), p. 259.

68. Ibid., p. 267.

69. Ibid., p. 260.

70. Ibid., pp. 265, 267.

71. David Horowitz, *The Professors: The 101 Most Dangerous Academics in America* (Washington, DC: Regnery, 2006).

72. Free Exchange on Campus, "Facts Count: An Analysis of David Horowitz's *The Professors: The 101 Most Dangerous Academics in America,*" May 2006, p.1.Online: http://www.freeexchangeoncampus.org/ index.php?option=com_docman&Itemid=25&task=view_category& catid=12&order=dmdate_published&ascdesc=DESC.

73. Horowitz, *The Professors,* p. 200.

74. Lewis R. Gordon, *Existentia Africana: Understanding Africana Existential Thought* (New York: Routledge, 2000), p. 4.

75. Cited in Bill Berkowitz, "Horowitz's Campus Jihads," *Dissident Voice* (October 9–19, 2004), pp. 1–6. Online: http://www.dissidentvoice.org/Oct04/Berkowitz1009.htm.

76. Cited in Leslie Rose, "David Horowitz: Battering Ram for Bush Regime," *Revolution Online* (August 28, 2005). Online: http://rwor.org/a/013/horowitz-battering-ram.htm.

77. Larissa Macfarquhar, "The Devil's Accountant," *New Yorker* (March 31, 2003). Online: http://www.analphilosopher.com/files/ MacFarquhar,_The_Devil's_Accountant_(2003).pdf.

78. David Horowitz, *Unholy Alliance: Radical Islam and the American Left* (New York: National Book Network, 2004), p. 56.

79. This silly shame and smear list can be found online: http://www. discoverthenetworks.com/individual.asp.

80. James Vanlandingham, "Capitol Bill Aims to Control Leftist Profs," *Independent Florida Alligator* (March 23, 2005). Online: http://www.alligator.org/2/050323freedom.php.

81. In the House of Representatives, the ABOR was taken up as HR 3077, which was part of HR 609. It is Title VI of the Higher Education Act. This is why it is also called Title VI in some discussions. The House version was also called the College Access and Opportunity Act and passed the House. It has been recommended with some significant revisions to the Senate as S 1614. For a summary of the differences, see the AAUP Web site: http://aaup.org/govrel/hea/index.htm.

82. Scott Jaschik, "$500 Fines for Political Profs," *Inside Higher Ed* (February 19, 2007). Online: http://insidehighered.com/layout/ set/print/news/2007/02/19/ariz.

83. Fish, "On Balance."

84. See Beinin, "The New McCarthyism."

85. I have taken up the issues of critical pedagogy, democracy, and schooling in a number of books. See Henry A. Giroux, *Border Crossings* (New York: Routledge, 2005); *Democracy on the Edge* (New York: Palgrave, 2006); *The Giroux Reader*, ed. Christopher Robbins (Boulder: Paradigm, 2006); Giroux and Giroux, *Take Back Higher Education*; and Giroux, *The University in Chains*.

86. Jacques Derrida, "The Future of the Profession or the Unconditional University," in *Derrida Down Under*, trans. Peggy Kamuf, ed. Laurence Simmons and Heather Worth (Palmerston North, NZ: Dunmore, 2001), p. 233.

87. Angus, "Academic Freedom in the Corporate University," pp. 67–68.

88. For an excellent analysis of contingent academic labor as part of the process of the subordination of higher education to the demands of capital and corporate power, see Marc Bousquet, *How the University Works: Higher Education and the Low-Wage Nation* (New York: New York University Press, 2008).

89. Angus, "Academic Freedom in the Corporate University," pp. 67–68.

90. These themes in Arendt's work are explored in detail in Elizabeth Young-Bruehl, *Why Arendt Matters* (New Haven: Yale University Press, 2006).

91. Sheldon S. Wolin, *Democracy, Inc.: Managed Democracy and the Specter of Inverted Totalitarianism* (Princeton: Princeton University Press, 2008), p. 147. It is worth noting that this is not an argument for not serving in the military, but a call for all citizens to serve, thereby sharing the military sacrifices a country has to make.

92. Jacques Rancière cited in Carnevale and Kelsey, "Art of the Possible," p. 263.

93. Hannah Arendt, *Origins of Totalitarianism* (New York: Harcourt Trade, 2001).

94. Angus, "Academic Freedom in the Corporate University," pp. 64–65.

95. On the relationship between education and hope, see Coté, Day, and de Peuter, eds., *Utopian Pedagogy*; and Henry A. Giroux, *Public Spaces/Private Lives: Democracy Beyond 9/11* (Boulder: Rowman and Littlefield, 2003).

96. I take up the issue of the increasing militarization of the university in Giroux, *The University in Chains*.

97. Wolin, *Democracy, Inc.*, p. 43.

98. Greig de Peuter, "Universities, Intellectuals, and Multitudes: An Interview with Stuart Hall," in *Utopian Pedagogy*, ed. Coté, Day, and de Peuter, pp. 113–114.

99. Zygmunt Bauman cited in Zygmunt Bauman and Keith Tester, *Conversations with Zygmunt Bauman* (London: Polity Press, 2001), p. 4.

100. Arundhati Roy, *Power Politics* (Cambridge, MA: South End Press, 2001), p. 6.

101. Wolin, *Democracy Inc.*, p. 161.

102. Richard J. Bernstein, *The Abuse of Evil: The Corruption of Politics and Religion since 9/11* (Malden: Polity Press, 2005), pp. 25–26.

103. Jacques Derrida, "Autoimmunity: Real and Symbolic Suicides—A Dialogue with Jacques Derrida," in *Philosophy in a Time of Terror: Dialogues with Jurgen Habermas and Jacques Derrida*, ed. Giovanna Borradori (Chicago: University of Chicago Press, 2004), p. 121.

104. Cited in Madeline Bunting, "Passion and Pessimism," *The Guardian* (April 5, 2003). Online: http:/books.guardian.co.uk/print/0,3858, 4640858,00.html.

105. Edward Said, *Humanism and Democratic Criticism* (New York: Columbia, 2004), p. 143.

106. Edward Said, "Scholarship and Commitment: An Introduction," *Profession* (2000), p. 7.

107. Kimball cited in Lawrence W. Levine, *The Opening of the American Mind* (Boston: Beacon Press, 1996), p. 19.

108. All of these ideas and the quotation itself are taken from Edward Said, *Humanism and Democratic Criticism*, p. 142.

109. Zygmunt Bauman, *Liquid Times: Living in an Age of Uncertainty* (London: Polity Press, 2007), p. 8.

110. Frederick Douglass, "The Significance of Emancipation in the West Indies, Speech, Canandaigua, New York, August 3, 1857," in *The Frederick Douglass Papers: Speeches, Debates, and Interviews*, Volume 3: 1855–63, ed. John W. Blassingame (New Haven: Yale University Press, 1985), p. 204.

## CHAPTER 4

1. Editorial, "Looking at America," *New York Times* (December 31, 2007), p. A20.

2. Ibid.

3. Sidney Blumenthal, "Bush's War on Professionals," *Salon.com* (January 5, 2006). Online: http://www.salon.com/opinion/ blumenthal/2006/01/05/spying/index. html.

4. Bob Herbert, "America the Fearful," *New York Times* (May 15, 2006), p. A25.

5. Jacques Rancière, *Hatred of Democracy* (London: Verso, 2006).

6. I have taken the term "torture factories" from Angela Y. Davis, *Abolition Democracy: Beyond Empire, Prisons, and Torture* (New York: Seven Stories Press, 2005), p. 50. The United States had 2,319,258 people in jail or prison at the start of 2008—one out of every hundred and more than any other nation. See Associated Press, "A First: 1 in 100 Americans Jailed," *MSNBC.com* (February 28, 2008). Online: http://www.msnbc.msn.com/id/23392251/print/1/ displaymode/1098/.

7. For an extensive treatment of this issue, see Michael McHugh, *The Second Gilded Age: The Great Reaction in the United States, 1973–2001* (Boulder: University Press, 2006). See also Mike Davis and Daniel Bertrand Monk, eds., *Evil Paradises* (New York: The New Press, 2007); Paul Krugman, "Gilded Once More," *New York Times* (May 27, 2007), online: http://select.nytimes.com/ 2007/04/27/opinion/27krugman.html?_r=1&hp&oref=slogin; Louis Uchitelle, "The Richest of the Rich, Proud of a New Gilded Age," *New York Times* (July 15, 2007), online: http://www. nytimes.com/2007/07/15/ business/15gilded.html?_r=1&oref=slogin; Peter Dreier, "Bush's Class Warfare," *CommonDreams.org* (December 22, 2007), online: http:// www.commondreams.org/archive/2007/12/22/5961/; and William Greider, "The Coming Political Revolution," *The Nation* (June 19, 2006), online: http://www.alternet.org/story/37359/; and Alan Trachtenberg, *The Incorporation of America: Culture and Society in the Gilded Age* (Vancouver: Douglas and McIntyre, 2007). See also the now classic Matthew Josephson, *The Robber Barons: The Great American Capitalists 1861–1901* (reprint) (New York: Harcourt, 2001).

8. Davis and Monk, "Introduction," *Evil Paradises*, p. ix.

9. For an extensive analysis of the First and Second Gilded Ages, see McHugh, *The Second Gilded Age.* While all of these factors connect the old and new Gilded Ages, one cannot but acknowledge the fact that both share a deep-seated racism, a Supreme Court deeply implicated in maintaining racist patterns of segregation, and a degree of poverty, violence, exclusion, and human suffering partly caused by the workings of a racist state. One important analysis between the Gilded Age and racism can be found in Susan Searls Giroux, "Race, Rhetoric, and the Contest Over Civic Education," in Henry A. Giroux and Susan Searls Giroux, *Take Back Higher Education* (New York: Palgrave Macmillan, 2006), pp. 129–167.

10. Paul Krugman, "Gilded Once More."

11. Mark Karlin, "Paul Krugman: Why Do Right-Wingers Mock Attempts to Care for Other People?" *AlterNet* (October 12, 2007). Online: http://www.alternet.org/workplace/64931/.

12. Krugman, "Gilded Once More."

13. Eric Lipton and Stephen Labaton, "Deregulator Looks Back, Unswayed," *New York Times* (November 17, 2008). Online: http://www.nytimes.com/2008/11/17/business/economy/17gramm.html?_r=1&hp=&oref=slogin&pagewanted=print.

14. Ibid.

15. As Lawrence Grossberg points out, neoliberalism "describes a political-economic project" whose "supporters are bound together by their fundamental opposition to Keynesian demand-side fiscal policy and to government regulation of business. Second, many neoliberals support laissez-faire and define the free economy as the absence of any regulation or control ... neoliberals tend to believe that, since the free market is the most rational and democratic system of choice, every domain of human life should be open to the forces of the marketplace. At the very least, that means that the government should stop providing services that would be better delivered by opening them up to the marketplace. Third, neoliberals believe that economic freedom is the necessary precondition for political freedom (democracy); they often act as if democracy were nothing but economic freedom or the freedom to choose. Finally neoliberals are radical individualists. Any appeal to larger groups (e.g., gender, racial, ethnic, or class groups) as if they functioned as agents or had rights, or to society itself, is not only meaningless but also a step toward socialism and totalitarianism." Lawrence Grossberg, *Caught in the Crossfire: Kids, Politics, and America's Future* (Boulder: Paradigm, 2005), p. 112. While the literature on neoliberalism is vast, some important examples include: Alain Touraine, *Beyond Neoliberalism* (London: Polity Press, 2001); Lisa Duggan, *The Twilight of Equality* (Boston: Beacon Press, 2003); David Harvey, *A Brief History of Neoliberalism* (New York: Oxford University Press, 2005); Wendy Brown, *Edgework: Critical Essays on Knowledge and Politics* (Princeton: Princeton University Press, 2005); Alfredo Saad-Filho and Deborah Johnston, eds., *Neoliberalism: A Critical Reader* (London: Pluto Press, 2005); Neil Smith, *The Endgame of Globalization* (New York: Routledge, 2005); Aihwa Ong, *Neoliberalism as Exception: Mutations in Citizenship and Sovereignty* (Durham: Duke University Press, 2006); Randy Martin, *An Empire of Indifference: American War and the Financial Logic of Risk Management* (Durham: Duke University Press, 2007); and Henry A. Giroux, *Against the Terror of Neoliberalism: Beyond the Politics of Greed* (Boulder: Paradigm, 2008).

16. For a thorough analysis of Bush's neoliberal belief that "markets do better when left alone," see Jo Becker, Sheryl Gay Stolberg, and Stephen

Labaton, "White House Philosophy Stoked Mortgage Bonfire," *New York Times* (December 21, 2008), pp. 1, 30–31. Also see, Linda J. Bilmes and Joseph E. Stiglitz, "The $10 Trillion Dollar Hangover: Paying the Price for Eight Years of Bush," *Harper's Magazine* (January 2009), pp. 31–35. For an illuminating article on the rise of neoliberalism in the 1990s, see David Kotz, "Neoliberalism and the U.S. Economic Expansion of the '90s," *Monthly Review* 54:11 (April 2003), p. 16.

17. Brown, *Edgework*, p. 42.

18. Ibid., pp. 40–41.

19. Ibid., p. 42.

20. The concept of the "market-state" is taken from neoconservative Philip Bobbitt's *The Shield of Achilles* (New York: Anchor, 2003). The quote is taken from Niall Ferguson, who supports Bobbitt's belief that the "market-state" is synonymous with democracy. See Niall Ferguson, "War Plans," *New York Times Book Review* (April 13, 2008), p. 10.

21. Brown, *Edgework*, p. 42.

22. See, especially, Naomi Klein, *The Shock Doctrine: The Rise of Disaster Capitalism* (Toronto: Alfred Knopf Canada, 2007). See also Kenneth J. Saltman, *Capitalizing on Disaster: Taking and Breaking Public Schools* (Boulder: Paradigm, 2007); Kenneth J. Saltman, ed., *Schooling and the Politics of Disaster* (New York: Routledge, 2007); and Lewis Gordon and Jane Gordon, *On Divine Warning: Disaster in the Modern Age* (Boulder: Paradigm, 2008).

23. Luis E. Carcamo-Huechante, "Milton Friedman: Knowledge, Public Culture, and Market Economy in the Chile of Pinochet," *Public Culture* 18:2 (Spring 2006), p. 414.

24. Brown, *Edgework*, p. 42.

25. Chris Hedges, "America Is in Need of a Moral Bailout," *Truthdig.com* (March 23, 2009). Online: http://www.truthdig.com/report/item/20090323_america_is_in_need_of_a_moral_bailout/.

26. Thomas Lemke, "Foucault, Governmentality, and Critique," *Rethinking Marxism* 14:3 (Fall 2002), p. 52.

27. Ibid., p. 50. On the issue of governmentality, see Michel Foucault, "Governmentality," trans. Rosi Braidotti and revised by Colin Gordon in *The Foucault Effect: Studies in Governmentality*, ed. Graham Burchell, Colin Gordon, and Peter Miller (Chicago: University of Chicago Press, 1991), pp. 87–104; Michel Foucault, "The Subject and the Power," in *Michel Foucault: Beyond Structuralism and Hermeneutics*, ed. Hubert Dreyfus and Paul Rabinow (Chicago: University of Chicago Press, 1983); and Michel Foucault, "Security, Territory and Population," in *Ethics: Subjectivity and Truth*, ed. Paul Rabinow (New York: The New Press, 1997), pp. 67–71.

28. I take up this issue in great detail in Giroux, *Against the Terror of Neoliberalism*.

29. Thomas Lemke, "A Zone of Indistinction: A Critique of Giorgio Agamben's Concept of Biopolitics," *Outlines: Critical Social Studies* 7:1 (2005), p. 12.

30. On the other hand, neoconservatives such as Niall Ferguson argue that the market state and democracy are self-identical precisely because both are "states of consent." Questions of who controls the means of consent or how inequality produces limited access to shaping the democratic nature of such apparatuses are lost on ideologues such as Ferguson. See Niall Ferguson, "War Plans," pp. 1, 10–11.

31. See Michel Foucault, *Society Must Be Defended: Lectures at the College De France 1975–1976*, trans. David Macey (New York: Picador, 2003); and Nikolas Rose, *The Politics of Life Itself: Biomedicine, Power, and Subjectivity in the Twenty-First Century* (Princeton: Princeton University Press, 2007).

32. Zygmunt Bauman, *The Individualized Society* (London: Polity Press, 2001), p. 9.

33. Catherine Needham, "Customer-Focused Government," *Soundings* 26 (Spring 2004), p. 80.

34. Ruth Rosen, "Note to Nancy Pelosi: Challenge Market Fundamentalism," *CommonDreams.org* (January 30, 2007). Online: http://www.commondreams.org/views07/0130–22.htm.

35. On the biopolitics of disposability, see Achille Mbembe, "Necropolitics," trans. Libby Meintjes, *Public Culture* 15:1 (2003), pp. 11–40; Zygmunt Bauman, *Wasted Lives* (London: Polity Press, 2004); Judith Butler, *Precarious Life: The Powers of Mourning and Violence* (London: Verso, 2004); and Arjun Appadurai, *Fear of Small Numbers: An Essay on the Geography of Anger* (Durham: Duke University Press, 2006).

36. Bruce Lambert and Christine Hauser, "Corpse Wheeled to Check-Cashing Store Leads to 2 Arrests ," *New York Times* (January 8, 2008). Online: http://www.nytimes.com/2008/01/09/nyregion/09dead.html?_r=1&ref=nyregion&oref=slogin.

37. Bob Herbert, "The Wrong Target," *New York Times* (February 19, 2008), p. A27.

38. Ben Zipperer, "Inmates vs. Animals: U.S. Fails the Test of Civilization," *AlterNet* (April 18, 2007). Online: http://www.alternet.org/rights/50428/.

39. Naomi Klein, *Fences and Windows* (New York: Picador, 2002), p. 21.

40. Martin, *An Empire of Indifference*, p. 139.

41. Ralph Nader, "Ralph Nader on the Candidates, Corporate Power and His Own Plans for 2008," *DemocracyNow.org* (July 9, 2007). Online: http://www.democracynow.org/article.pl?sid=07/07/09/131226.

42. Martin, *An Empire of Indifference*, p. 138.

43. I want to make it clear that I am not arguing that state sovereignty in the past represented some ideal or "good" form of sovereignty. State sovereignty is inscribed in a damaged legacy of social, economic,

and racial exclusion. But this should not suggest that state sovereignty does not function differently in various historical conjunctures, sometimes more and sometimes less on the side of justice. Nor should my comments suggest that sovereignty is not worth fighting over, even if it means gaining incremental modes of justice while refusing to embrace liberal social democratic aims. When sovereignty is connected with the concept of democracy, it contains the possibility of being contested. Corporate and state sovereignty under neoliberalism seem incapable of questioning and contesting themselves and close off matters of debate, reform, and transformation, if not politics itself, through endless hegemonic claims "that presuppose [their] own perfectability, and thus [their] own historicity." The quote is from Jacques Derrida, "Autoimmunity: Real and Symbolic Suicides—A Dialogue with Jacques Derrida," in *Philosophy in a Time of Terror: Dialogues with Jurgen Habermas and Jacques Derrida*, ed. Giovanna Borradori (Chicago: University of Chicago Press, 2004), p. 121.

44. Richard J. Bernstein, *The Abuse of Evil: The Corruption of Politics and Religion since 9/11* (London: Polity Press, 2005), p. 76.
45. David R. Francis, "What A New 'Gilded Age' May Bring," *Christian Science Monitor* (March 8, 2006). Online: http://www.csmonitor.com/2006/0306/p16s01-coop.html.
46. Kirk Johnson, "As Logging Fades, Rich Carve Up Open Land in West," *New York Times* (October 13, 2007), p. A1.
47. Uchitelle, "The Richest of the Rich, Proud of a New Gilded Age."
48. Christine Haughney, "It's Not So Easy Being Less Rich," *New York Times* (June 1, 2007), p. 10.
49. Ruth La Ferla, "Markets Stall but Spoiled Always Sells," *New York Times* (October 23, 2008), pp. E1, E8.
50. Bauman, *Wasted Lives* (London: Polity Press, 2004), p. 131.
51. Francis, "What A New 'Gilded Age' May Bring."
52. Uchitelle, "The Richest of the Rich, Proud of New Gilded Age."
53. Robert Reich, "Obama, Bitterness, Meet the Press, and the Old Politics," *CommonDreams.org* (April 15, 2008). Online: http://www.commondreams.org/archive/2008/04/15/8300/.
54. Christine Haughney and Eric Konigsberg, "Even When Times Get Tough, the Ultrarich Keep Spending," *New York Times* (April 14, 2007), pp. A1, A17.
55. Davis and Monk, "Introduction," *Evil Paradises*, pp. ix–x.
56. Dollars and Sense and United for a Fair Economy, eds., *The Wealth Inequality Reader*, 2nd edition (Chicago: Dollars and Sense, 2008).
57. Robert Weissman, "The Shameful State of the Union," *Huffington Post* (January 29, 2008). Online: http://www.huffingtonpost.com/ robert-weissman/the-shameful-state-of-the_b_83844.html.
58. Peter Dreier, "Bush's Class Warfare." See also Chris Hartman, "By the Numbers," *Inequality.org* (October 14, 2007). Online:

http://www.demos.org/inequality/numbers.cfm. As David R. Francis points out, "the richest of the rich, the top 1/1,000th, enjoyed a 497 percent gain in wage and salary income between 1972 and 2001. Those at the 99th percentile, who made an average $1.7 million per year in 2001, enjoyed a mere 181 percent gain." See Francis, "What A New 'Gilded Age' May Bring."

59. Ellen Simon, "Half of S&P 500 CEOs Topped $8.3 Million," *Associated Press* (June 11, 2007). Online: http://www.washingtonpost. com/wp-dyn/content/article/2007/06/11/AR2007061100798.html.

60. Ibid.

61. Dean Baker, "Year of the Fat Cats," *CommonDreams.org* (January 17, 2008). Online: http://www.commondreams.org/archive/2008/01/17/6420/.

62. Jenny Anderson, "Chiefs' Pay Under Fire at Capitol," *New York Times* (March 8, 2008), p. B4.

63. Ibid.

64. Paul Krugman, "Money for Nothing," *New York Times* (April 27, 2009), P. A21.

65. Edmund L. Andrews, "Bush Vows to Veto a Mortgage Relief Bill," *New York Times* (February 27, 2008). Online: http://www.nytimes. com/2008/02/27/business/27housing.html?sq=Edmund%20L.%20 Andrews&st=nyt&scp=4&pagewanted=print.

66. Edmund L. Andrews, "Partisan Split Emerges on Mortgage Crisis Relief," *New York Times* (February 29, 2008), p. C1.

67. Cited in Ruth Conniff, "Drowning the Beast," *The Progressive* (September 7, 2005). Online: http://www.commondreams.org/views05/0907-30.htm.

68. Paul Krugman, "Conservatives Are Such Jokers," *New York Times* (October 5, 2007), p. A27.

69. Angela Davis, "Locked Up: Racism in the Era of Neoliberalism," *ABC News Opinion* (Australia) (March 18, 2008). Online: http://www.abc.net.au/news/stories/2008/03/19/2193689.htm.

70. I take up this issue in great detail in Henry A. Giroux, *The University in Chains: Confronting the Military-Industrial-Academic Complex* (Boulder: Paradigm, 2007).

71. Appadurai, *Fear of Small Numbers*, pp. 36–37.

72. For an excellent set of essays on the new extremes of wealth and power, see Dollars and Sense and United for a Fair Economy, eds., *The Wealth Inequality Reader*.

73. Nick Couldry, "Reality TV, or the Secret Theater of Neoliberalism," *Review of Education, Pedagogy, and Cultural Studies* 30:1 (January–March 2008), pp. 3–13.

74. Davis and Monk, "Introduction," *Evil Paradises*, p. ix.

75. Associated Press, "A First: 1 in 100 Americans Jailed."

76. Harvey, *A Brief History of Neoliberalism*, p. 7.

77. Ibid., p. 161.

78. Zygmunt Bauman, *Liquid Times: Living in an Age of Uncertainty* (London: Polity Press, 2007), p. 28.

79. Thom Hartmann, "You Can't Govern If You Don't Believe in Government," *CommonDreams.org* (September 6, 2005). Online: http://www.commondreams.org/views05/0906-21.htm.

80. Some of the most brilliant work on racist exclusion can be found in David Theo Goldberg, *Racist Culture* (Malden, MA: Blackwell, 1993); and David Theo Goldberg, *The Threat of Race: Reflections on Racial Neoliberalism* (Malden, MA: Blackwell, 2009).

81. Zygmunt Bauman, *Consuming Life* (London: Polity Press, 2007), p. 65.

82. Mbembe, "Necropolitics," pp. 11–12.

83. Ewa Ponowska Ziarek, "Bare Life on Strike: Notes on the Biopolitics of Race and Gender," *South Atlantic Quarterly* 107:1 (Winter 2008), p. 90.

84. See Michel Foucault, *The History of Sexuality. An Introduction* (New York: Vintage Books, 1990); and Foucault, *Society Must Be Defended*.

85. Mika Ojakangas, "Impossible Dialogue on Bio-power: Agamben and Foucault," *Foucault Studies* 2 (May 2005), pp. 5–6.

86. Foucault, *Society Must Be Defended*, p. 249.

87. Foucault, *History of Sexuality*, p. 137.

88. Lemke, "A Zone of Indistinction," p. 11.

89. Foucault, "Governmentality," p. 103.

90. Ojakangas, "Impossible Dialogue on Bio-Power," p. 6.

91. Foucault, *Society Must Be Defended*, p. 255.

92. Lemke, "Foucault, Governmentality, and Critique," p. 50.

93. Lemke, "A Zone of Indistinction," p. 3.

94. Butler, *Precarious Life*, p. 52.

95. Giorgio Agamben, *Homo Sacer: Sovereign Power and Bare Life*, trans. Daniel Heller-Roazen (Stanford: Stanford University Press, 1998), p. 4.

96. For a brilliant analysis of the differences between Agamben's and Foucault's concept of biopolitics, see Lemke, "A Zone of Indistinction."

97. See, especially, Giorgio Agamben, *Homo Sacer*; Giorgio Agamben, *Remnants of Auschwitz: The Witness and the Archive*, trans. Daniel Heller-Roazen (Cambridge: Zone Books, 2002); and Giorgio Agamben, *State of Exception*, trans. Kevin Attell (Chicago: University of Chicago, 2003).

98. For an extensive elaboration of Agamben's notion of the camp as the nomos of society, see Bulent Diken and Carsten Bagge Lausten, *The Culture of Exception: Sociology Facing the Camp* (New York: Routledge, 2005). I would argue that this work is very suggestive, but it fails to adequately differentiate important political and theoretical differences between Agamben and other theorists of biopower and its view of the camp is too overdetermined.

99. Agamben, *Homo Sacer*, p. 166.

100. Ibid., pp.168–169, emphasis in original.

101. See, especially, Giorgio Agamben, *Homo Sacer*, p. 8.
102. Ibid., p. 181.
103. Zygmunt Bauman, *Liquid Love* (London: Polity Press, 2003), p. 133.
104. Malcolm Bull, "States Don't Really Mind Their Citizens Dying (Provided They Don't All Do It at Once): They Just Don't Like Anyone Else to Kill Them," *London Review of Books* (December 16, 2004), p. 3.
105. Agamben, *Homo Sacer*, p. 166.
106. Lemke, "A Zone of Indistinction," p. 6.
107. Catherine Mills, "Agamben's Messianic Biopolitics: Biopolitics, Abandonment and Happy Life," *Contretemps* 5 (December 2004), p. 47.
108. Agamben, *Homo Sacer*, p. 165.
109. Mills, "Agamben's Messianic Biopolitics," p. 47.
110. Jean Comaroff, "Beyond Bare Life: AIDS, (Bio)Politics, and the Neoliberal Order," *Public Culture* 19:1 (Winter 2007), p. 209.
111. Lemke, "A Zone of Indistinction," p. 2.
112. The following section draws from Henry A. Giroux, *Stormy Weather: Katrina and the Politics of Disposability* (Boulder: Paradigm, 2006).
113. Rosa Brooks, "Our Homegrown Third World," *Los Angeles Times* (September 7, 2005), pp. 1–2. Online: http://www.commondreams.org/cgi-bin/print.cgi?file=/views05/0907-24.htm.
114. Angela Davis, *Abolition Democracy*, pp. 122, 124.
115. Of course, the social contract largely excluded black populations in the United States. See David Theo Goldberg, *The Racial State* (Malden, MA: Blackwell, 2002). I want to make it clear that I am not invoking some idealized notion of the social contract, which in the past has been damaged and iniquitous, but, rather, invoking the promise of a social contract that not only makes the issue of social provisions and guarantees partly a matter of state responsibility but also inscribes a social bond that, while always impossible, is central to any struggle for democracy.
116. Agamben, *Homo Sacer*, p. 9.
117. William J. Cromie, "AIDS Epidemic Called Crisis Among Blacks," *Harvard University Gazette* (March 19, 1998). Online: http://www.hno.harvard.edu/gazette/1998/03.19/AIDSEpidemicCal.html.
118. Cited from ABC's special edition of *Primetime*, which also reported that "Black Americans make up 13% of the U.S. population but account for over 50% of all new cases of HIV, the virus that causes AIDS. That infection rate is eight times the rate of whites. Among women, the numbers are even more shocking—almost 70% of all newly diagnosed HIV positive women in the U.S. are Black women. Black women are 23 times more likely to be diagnosed with AIDS than white women, with heterosexual contact being the overwhelming method of infection in Black America." See Transcript, "The Silent Killer in the US Out of Control: AIDS in Black America," special edition of *Primetime* on ABC (August 24, 2006). Online: http://www.thefutoncritic.com/listings.aspx?id=20060810abc02.

119. Susy Buchanan and David Holthouse, "Locked and Loaded," *The Nation* (August 28/September 4, 2006), pp. 29–32.

120. Samuel P. Huntington, *Who Are We? The Challenges to America's National Identity* (New York: Simon & Schuster, 2004). For an excellent review of the book, see Louis Menand, "Patriot Games: The New Nativism of Samuel P. Huntington," *New Yorker* (May 17, 2004), pp. 92–98.

121. Editorial, "Gitmos Across America," *New York Times* (June 27, 2007), p. A22. Online: http://www.nytimes.com/2007/06/27/opinion/27weds1.html.

122. Nikhil Singh, "The Afterlife of Fascism," *South Atlantic Quarterly* 105:1 (Winter 2006), pp. 83–84.

123. Christopher Moraff, "America's Slave Labor," *In These Times* (January 17, 2007), pp. 1–3. Online: http://www.inthesetimes.com/article/2982/.

124. Southern Poverty Law Center, *Close to Slavery: Guestworker Programs in the United States* (Montgomery, AL: The Southern Poverty Law Center, 2007). Online: http://www.splcenter.org/pdf/static/ SPLCguestworker.pdf.

125. Pat Eaton-Robb, "Homeless Families on the Rise, with No End in Sight," *Associated Press* (October 8, 2007). Online: http://www.commondreams.org/archive/2007/10/08/4403/.

126. Mitchell Dean, "Four Theses on the Powers of Life and Death," *Contretemps* 5 (December 2004), p. 17.

127. Lemke, "A Zone of Indistinction," p. 10.

128. Ibid., p. 10.

129. Ernesto Laclau, "Bare Life or Social Indeterminancy?" in *Giorgio Agamben: Sovereignty and Life*, ed. Matthew Calauco and Steven De Caroli (Stanford: Stanford University Press, 2007), pp. 13–14.

130. Comaroff, "Beyond Bare Life," p. 210.

131. Ziarek, "Bare Life on Strike," p. 93.

132. Laclau, "Bare Life or Social Indeterminancy?" p. 22.

133. Cornelius Castoriadis, *Figures of the Thinkable*, trans. Helen Arnold (Stanford: Stanford University Press, 2007), p. 47.

134. Sheldon Wolin, "Political Theory: From Vocation to Invocation," in *Vocations of Political Theory*, ed. Jason Frank and John Tambornino (Minneapolis: University of Minnesota Press, 2000), p. 3.

135. Mbembe, "Necropolitics," pp. 39–40.

136. Sheldon Wolin, "Political Theory: From Vocation to Invocation," p. 15.

137. In addition to works cited above by Foucault and Agamben, see Michael Hardt and Antonio Negri, *Empire* (Cambridge: Harvard University Press, 2000); and Michael Hardt and Antonio Negri, *Multitude: War and Democracy in the Age of Empire* (New York: Penguin, 2004).

138. See Michel Foucault, "Afterword," in *Power/Knowledge: Selected Interviews and Other Writings 1972–1977*, ed. Colin Gordon (New York: Pantheon, 1980), especially pp. 208–226.

139. Foucault, *History of Sexuality*, p. 136.
140. Foucault, *Society Must Be Defended*, p. 255.
141. Michel Foucault, "About the Beginning of the Hermeneutics of the Self," *Political Theory* 21:2 (May 1993), pp. 203–204.
142. Foucault, "Governmentality," pp. 87–104.
143. Lemke, "Foucault, Governmentality, and Critique," p. 55.
144. Hardt and Negri, *Empire*, p. 23.
145. Ibid., pp. xiv–xv.
146. Hardt and Negri, *Multitude*, p. 67.
147. Ibid., p. 334.
148. Davis, "Locked Up: Racism in the Era of Neoliberalism."
149. Martin, *An Empire of Indifference*, p. 20.
150. Dollars and Sense and United for a Fair Economy, eds., *The Wealth Inequality Reader*.
151. I am drawing from a number of works by Zygmunt Bauman: *Liquid Times*; *Liquid Fear* (London: Polity Press, 2006); *Liquid Life* (London: Polity Press, 2005); and, especially, *Wasted Lives*.
152. Bauman, *Liquid Times*, pp. 45–46.
153. For a summary of a variety of indexes measuring the state of American democracy, see Executive Summary, *What Together We Can Do: A Forty Year Update of the National Advisory Commission on Civil Disorders* (Washington, DC: Eisenhower Foundation, 2008). Online: http://www.eisenhowerfoundation.org/docs/Kerner%2040%20Year%20Update,%20Executive%20Summary.pdf.
154. Klein, *The Shock Doctrine*.
155. Zygmunt Bauman, *Liquid Love* (London: Polity Press, 2003), p. 138.
156. Michael Dillon, " 'Sovereignty and Governmentality': From the Problematics of the 'New World Order' to the Ethical Problematic of the World Order," *Alternatives* 20 (1995), p. 330.
157. See Eva Illouz, *Cold Intimacies: The Making of Emotional Capitalism* (London: Polity Press, 2007).
158. John Dewey, *Democracy and Education* (orig. 1916) (New York: Macmillan, 1966). Dewey's position on public life, education, and democracy does not suggest an uncritical endorsement of his theory of pragmatism. For an interesting critique of Dewey's pragmatism, see Stanley Aronowitz, "Introduction," in *Critical Theory: Selected Essays by Max Horkheimer* (New York: Continuum, 1999), pp. xi–xxi.
159. John Dewey, *The Public and Its Problems* (New York: Henry Holt, 1927).
160. Hannah Arendt, *Between Past and Future* (New York: Penguin Books, 1977), p. 241.
161. Hannah Arendt, *Totalitarianism: Part Three of the Origins of Totalitarianism* (New York: Harcourt, 1976), p. 162.
162. See, especially, Hannah Arendt, *The Origins of Totalitarianism*, 3rd edition, revised (New York: Harcourt Brace Jovanovich, 1968); and

John Dewey, *Liberalism and Social Action* (orig. 1935) (New York: Prometheus Press, 1999).

163. Comaroff, "Beyond Bare Life," p. 207.
164. See, especially, Cornelius Castoriadis, "The Greek Polis and the Creation of Democracy," in *Philosophy, Politics, Autonomy: Essays in Political Philosophy* (New York: Oxford University Press, 1991), pp. 81–123.
165. Comaroff, "Beyond Bare Life," p. 208.
166. Martin, *An Empire of Indifference*, p. 141.
167. Comaroff, "Beyond Bare Life," p. 209.
168. Roger I. Simon, *The Touch of the Past* (New York: Palgrave, 2005), p. 117.
169. Bauman, *Liquid Life*, p. 124.
170. Ibid., p. 126.
171. Theodor W. Adorno, *Critical Models: Interventions and Catchwords*, trans. Henry W. Pickford (New York: Columbia University Press, 1998), p. 14.
172. Comaroff, "Beyond Bare Life," p. 200.
173. Ibid., p. 214.
174. Nick Couldry, "In Place of a Common Culture, What?" *Review of Education, Pedagogy, and Cultural Studies* 26:1 (2004), p. 15. On the issue of emergent publics, see Ian Angus, *Emergent Publics* (Winnipeg, Canada: Arbeiter Ring, 2001).
175. I am drawing here from Derrida, "Autoimmunity," pp. 85–136.

# INDEX